Legume Crops – Characterization and Breeding for Improved Food Security

Edited by Mohamed Ahmed El-Esawi

Published in London, United Kingdom

IntechOpen

Supporting open minds since 2005

Legume Crops - Characterization and Breeding for Improved Food Security
http://dx.doi.org/10.5772/intechopen.73753
Edited by Mohamed Ahmed El-Esawi

Contributors
Renée Fortunato, Virginia Fuentes Baluzzi, Fernando De Diego, Rodrigo Biagioni, Alejandro Esquivel, Faisal Hussain, Farzana Usman, Jolanta Bojarszczuk, Jerzy Księżak, Suhel Mehandi, Indra Prakash Singh, Sudhakar Prasad Mishra, Syed Mohd. Quatadah, Nagmi Praveen, Namrata Dwivedi, Ana Ribeiro De Barros, Ana Gomes, Nascimento Nhantumbo, Rafael Massinga, José Ramalho, Joyce Aparecida Tavares De Miranda, Lucia Maria Jaeger De Carvalho, Ana Vieira, Izabela Miranda De Castro, André Guimarães, Jose Carvalho, W. James Grichar, Peter A. Dotray, Mark Matocha, Mohamed A. El-Esawi

First published in London, United Kingdom, 2019 by IntechOpen
IntechOpen is the global imprint of INTECHOPEN LIMITED, registered in England and Wales, registration number: 11086078, 7th floor, 10 Lower Thames Street, London, EC3R 6AF, United Kingdom
Printed in Croatia

British Library Cataloguing-in-Publication Data
A catalogue record for this book is available from the British Library

Additional hard and PDF copies can be obtained from orders@intechopen.com

Legume Crops - Characterization and Breeding for Improved Food Security
Edited by Mohamed Ahmed El-Esawi
p. cm.
Print ISBN 978-1-83968-086-1
Online ISBN 978-1-83968-087-8
eBook (PDF) ISBN 978-1-83968-088-5

For EU product safety concerns:
IN TECH d.o.o., Prolaz Marije Krucifikse Kozulić 3, 51000 Rijeka, Croatia, info@intechopen.com or visit our website at intechopen.com.

Meet the editor

Dr. Mohamed Ahmed El-Esawi is a visiting research fellow at the University of Cambridge in the United Kingdom, and an associate professor of molecular genetics at the Botany Department of Tanta University in Egypt. Dr. El-Esawi received his BSc and MSc from Tanta University, and his PhD degree in Plant Genetics and Molecular Biology from Dublin Institute of Technology, Technological University Dublin in Ireland. Afterwards, Dr. El-Esawi joined the University of Warwick in the United Kingdom, University of Sorbonne (Paris VI) in France, and University of Leuven (KU Leuven) in Belgium as a visiting research fellow. His research focuses on plant genetics, genomics, molecular biology, molecular physiology, developmental biology, plant–microbe interaction, and bioinformatics. He has authored several international journal articles and book chapters, and participated in more than 60 conferences and workshops worldwide. Dr. El-Esawi has several awards and is currently involved in several research projects on biological sciences.

Contents

Preface

Legumes are flowering plants found in most of the archeological records of plants. Legumes are efficiently used as food crops for humans and animals, pulps for paper and timber manufacturing, sources for fuel and oil production, ornamental plants, and cover crops such as cereals and other staple foods. Additionally, they can be utilized for other purposes, including the production of massive amounts of organic nitrogen. This book reviews the fundamental advances related to the characterization and breeding of legume crops for improved food security. Moreover, it sheds new light on the current research trends and future research directions related to legume crop studies. This book will provoke interest for various readers, researchers, and scientists, who may find this information useful for the advancement of legume productivity.

The book includes eight chapters. The first introductory chapter "Characterization and improvement of legume crops" presents an introduction to the main legumes and enhancement of their productivity. The second chapter "Novel therapeutic uses of legume crops in southern South America" reviews some new therapeutic uses of legumes in southern South America. The third chapter "Ethnomedicinal values of legume plants in Pakistan" highlights the knowledge and importance of medicinal flora as well as traditional uses of such plants in daily life in Pakistan. The fourth chapter "Starches granules from cowpea, black and carioca beans in raw and cooked forms" evaluates the structure of common bean starch granules and cowpea in raw and cooked forms by optical microscopy and scanning electron microscopy. The fifth chapter "Mungbean (*Vigna radiata* L. Wilczek): retrospect and prospects" overviews mungbean crops and their retrospect and prospects. The sixth chapter "The productivity of selected species and cultivars of legumes grown for seeds in organic production systems" assesses the yielding of selected legume species with diversified morphological structure cultivated for seeds in ecological systems. The seventh chapter addresses the influence of adjuvants on the efficacy of postemergence herbicides commonly used in peanut (*Arachis hypogaea* L.). The eighth chapter studies breeding of elite cowpea (*Vigna unguiculata* (L.) Walp) varieties for improved food security and income in Africa.

The book editor would like to thank Mr. Edi Lipović, Author Service Manager, for his wholehearted cooperation in the publication of this book.

Mohamed Ahmed El-Esawi, PhD
Sainsbury Laboratory,
University of Cambridge,
Cambridge, United Kingdom

Botany Department,
Faculty of Science,
Tanta University,
Egypt

Chapter 1

Introductory Chapter: Characterization and Improvement of Legume Crops

Mohamed A. El-Esawi

1. Introduction

Legumes are agriculturally grown flowering plants that are found in most of the archaeological record of plants [1]. Various ecosystems, including rain forests, arctic/alpine regions, and deserts have been colonized by legumes [1, 2]. The most two popular flowering plants are Asteraceae and Orchidaceae [1]. The third in terms of popularity is Leguminosae or Fabaceae with 670–750 genera and 18,000–19,000 species, respectively [1]. Legumes are utilized efficiently as (a) food crops for humans and animals; (b) pulps for paper, wood, and timber manufacturing; (c) sources for fuel and oil production; (d) ornamental plants used as living barriers and firebreaks, among others [3]; and (e) cover crops such as cereals and other staple foods [1]. Additionally, they can be utilized for other purposes including production of massive amounts of organic nitrogen. This is because legumes can be intercropped with rhizobia resulting in high yield productivity, soil organic matter improvement, modification of soil osmosis and texture, nutrient reuse, decrease of soil pH and soil pressure, microorganism differentiation, and alleviation of disease problems [1, 4]. Furthermore, legumes can produce amounts of organic nitrogen at a slow rate when rotated with cereals. Such nitrogen produced can be utilized in prospective cropping technologies for improving the production of these crops, recognizing their potential role in promoting better human nutrition and soil health [1, 5].

2. Main legumes

Forage legumes such as alfalfa (*Medicago sativa*), clover (*Trifolium* spp.), bird's-foot trefoil (*Lotus corniculatus*), and vetch (*Vicia* spp.) are utilized as main sources for dairy and meat which are used for protein, fiber, and energy production [1]. Global production of alfalfa was approximately 436 tons in 2006 suggesting that it is the most essential forage crop. The highest amount of alfalfa was produced in the United States, being produced around 15 million tons in 2010 [1, 6]. Grain legumes or pulses are crops harvested massively for the dry seeds. They are found containing high amounts of protein in their seeds. Therefore, they represent a major food source for population consumption. They are considered as the main protein suppliers especially for people from developing countries [1]. Additionally, their high amino acid content is of nutritional value during utilization of cereals and tubers as food sources [1, 7]. The soybean (*Glycine max*), a native plant of Eastern Asia, is an annual summer legume of great agricultural possibilities due to its fundamental role in the nutrition of many people and livestocks besides its industrial possibilities [1, 8].

3. Enhancing legume productivity

Legumes are highly diversified, so they are utilized for several economic and cultural purposes including their role as vegetables to tolerate various ecological conditions, their source for producing large quantities of proteins, their utilization in grazing domains, and their function in increasing worldwide productivity of food and other commodities [1]. Therefore, recent findings have directed towards developing new biological and environment-friendly techniques to enhance the growth efficiency of legumes [1]. Scientists have derived several economic and ecological uses when legumes form symbiotic associations with nitrogen-fixing fungi and bacteria [9]. Biological nitrogen fixation (BNF) within legumes occurs through their association with microorganisms [1]. These microorganisms, which are also needed for the Earth's nitrogen cycle, are utilized for developing agricultural production of plants. Furthermore, they participate in soil colonization and plant growth promotion when utilized in live formulations or biofertilizers applied to seed, root, soil, or the interior of the plant as they can supply large amount of proteins to host cell and enhance soil protection [1]. The need of agroecosystems for nitrogen is assessed through a cost-effective, prospective, and eco-friendly process of biological nitrogen fixation rather than chemical nitrogen fixation. There are several benefits to the process of biological nitrogen fixation. It meets the needs of legumes and intercropped or succeeding crops for nitrogen [1]. This, in turns, avoids or even restricts the application of nitrogen fertilization. Additionally, nitrogen-fixing organisms play an essential role when the amount of nitrogen in the soil is low. They introduce ammonium into the legume biomass to allow faster growing than their plant competitors, but if the protein content is high, nitrogen-fixing microorganisms become alternative to non-fixing species due to high bioenergy cost of nitrogen fixation process [1, 10]. Thus, it can be concluded that nitrogen fixation in legume systems occurs through a variety of physiological and ecological possibilities including the plant's need for nitrogen and the C:N stoichiometry of the ecosystem [1]. It has proven experimentally and theoretically the hypothesis of a feedback control between legume's need for nitrogen and BNF in a specific ecosystem [11].

To enhance the efficiency of the nitrogen fixation process, the most suitable microorganisms for such purpose are selected, and/or genetic engineering of plant species are involved to guarantee high legume crop productivity [1]. Farmers are familiar with the application of commercially available microorganisms (inoculants) that are of great efficiency to nodulate plants and fix nitrogen in the soil [1]. These microorganisms such as rhizobia form associations with legumes in a situation called symbiosis that introduces benefits for both parts [1]. In this scenario, leguminous plants represent the source of energy and photosynthetic products to rhizobia, while rhizobia supplies the legumes with nitrogen in form of ammonium [1, 12]. The symbiosis begins when the roots of leguminous plants are inoculated the rhizobia, which, in turn, form root nodules where BNF occurs with the help of nitrogenase enzyme [1, 13]. In conclusions, several techniques have been developed genetically and biochemically to enhance plant development and crop productivity, suggesting their marvelous importance in improving legumes and other crops [1, 14–37].

Author details

Mohamed A. El-Esawi
Botany Department, Faculty of Science, Tanta University, Tanta, Egypt

*Address all correspondence to: mohamed.elesawi@science.tanta.edu.eg

References

[1] Morel MA, Braña V, Castro-Sowinski S. Legume Crops, Importance and Use of Bacterial Inoculation to Increase Production, Crop Plant, Aakash Goyal. IntechOpen; 2012. DOI: 10.5772/37413. Available from: https://www.intechopen.com/books/crop-plant/legume-crops-importance-and-use-of-bacterial-inoculation-to-increase-production

[2] Schrire BD, Lewis GP, Lavin M. Biogeography of the Leguminosae. In: Lewis G, Schrire G, Mackinder B, Lock M, editors. Legumes of the World. Kew, UK: Royal Botanic Gardens. ISBN: 8773043044; 2005. pp. 21-54

[3] Lewis G, Schrire B, MacKinder B, Lock M. Legumes of the World. UK: Royal Botanical Gardens, Kew Publishing; 2005. ISBN: 1900347806

[4] USDA. Soil Quality-Agronomy Technical Note. 1998. Available from: http://soils.usda.gov/sqi/management/files/sq_atn_6.pdf [Accessed: 30 August 2011]

[5] Popelka C, Terryn N, Higgins T. Gene technology for grain legumes: Can it contribute to the food challenge in developing countries? Plant Science. 2004;**167**:195-206. ISSN: 01689452

[6] FAO STAT. Food and Agriculture Organization Statistical Database. 2011. Available from: http://faostat.fao.org [Accessed: 30 August]

[7] Graham PH, Vance CP. Legumes: Importance and constraints to greater use. Plant Physiology. 2003;**131**:872-877. ISSN: 0032-0889

[8] Ali N. Soybean processing and utilization. In: Singh G, editor. The Soybean: Botany, Production and Uses. UK: CAB International. ISBN: 9781845936440; 2010. pp. 345-362

[9] Zahran HH. Enhancement of Rhizobia-legumes symbioses and nitrogen fixation for crops productivity improvement. In: Khan MS, Zaidi A, Musarrat J, editors. Microbial Strategies for Crop Improvement. Netherlands: Springer; 2009. pp. 227-254. ISBN: 9783642019784

[10] Houlton BZ, Wang YP, Vitousek PM, Field CB. A unifying framework for dinitrogen fixation in the terrestrial biosphere. Nature. 2008;**454**:27-334. ISSN: 0028-0836

[11] Soussana J-F, Tallec T. Can we understand and predict the regulation of biological N_2 fixation in grassland ecosystems? Nutrient Cycling in Agroecosystems. 2010;**88**:197-213. ISSN: 1385-1314

[12] Howard JB, Rees DC. Structural basis of biological nitrogen fixation. Chemical Reviews. 1996;**96**:2965-2982. ISSN: 0009-2665

[13] Masson-Boivin C, Giraud E, Perret X, Batut J. Establishing nitrogen-fixing symbiosis with legumes: How many *Rhizobium* recipes? Trends in Microbiology. 2009;**17**:458-466. ISSN: 0966-842X

[14] Elansary HO, Yessoufou K, Abdel-Hamid AME, El-Esawi MA, Ali HM, Elshikh MS. Seaweed extracts enhance Salam Turfgrass performance during prolonged irrigation intervals and saline shock. Frontiers in Plant Science. 2017;**8**:830

[15] El-Esawi MA. Micropropagation technology and its applications for crop improvement. In: Anis M, Ahmad N, editors. Plant Tissue Culture: Propagation, Conservation and Crop Improvement. Singapore: Springer; 2016. pp. 523-545

[16] El-Esawi MA. Nonzygotic embryogenesis for plant development.

In: Anis M, Ahmad N, editors. Plant Tissue Culture: Propagation, Conservation and Crop Improvement. Singapore: Springer; 2016. pp. 583-598

[17] El-Esawi MA. Somatic hybridization and microspore culture in brassica improvement. In: Anis M, Ahmad N, editors. Plant Tissue Culture: Propagation, Conservation and Crop Improvement. Singapore: Springer; 2016. pp. 599-609

[18] El-Esawi MA. Genetic diversity and evolution of *Brassica* genetic resources: From morphology to novel genomic technologies—A review. Plant Genetic Resources. 2017;**15**:388-399

[19] El-Esawi MA. SSR analysis of genetic diversity and structure of the germplasm of faba bean (*Vicia faba* L.). Comptes Rendus Biologies. 2017;**340**:474-480

[20] El-Esawi MA, Alayafi AA. Overexpression of rice *Rab7* gene improves drought and heat tolerance and increases grain yield in rice (*Oryza sativa* L.). Genes. 2019;**10**:56

[21] El-Esawi MA, Sammour R. Karyological and phylogenetic studies in the genus *Lactuca* L. (Asteraceae). Cytologia. 2014;**79**:269-275

[22] El-Esawi MA, Al-Ghamdi AA, Ali HM, Alayafi AA, Witczak J, Ahmad M. Analysis of genetic variation and enhancement of salt tolerance in French pea (*Pisum sativum* L.). International Journal of Molecular Sciences. 2018;**19**:2433

[23] El-Esawi MA, Alaraidh IA, Alsahli AA, Ali HM, Alayafi AA, Witczak J, et al. Genetic variation and alleviation of salinity stress in barley (*Hordeum vulgare* L.). Molecules. 2018;**23**:2488

[24] El-Esawi MA, Alaraidh IA, Alsahli AA, Alamri SA, Ali HM, Alayafi AA. *Bacillus firmus* (SW5) augments salt tolerance in soybean (*Glycine max* L.) by modulating root system architecture, antioxidant defense systems and stress-responsive genes expression. Plant Physiology and Biochemistry. 2018;**132**:375-384

[25] El-Esawi MA, Alaraidh IA, Alsahli AA, Alzahrani SM, Ali HM, Alayafi AA, et al. *Serratia liquefaciens* KM4 improves salt stress tolerance in maize by regulating redox potential, ion homeostasis, leaf gas exchange and stress-related gene expression. International Journal of Molecular Sciences. 2018;**19**:3310

[26] El-Esawi MA, Elansary HO, El-Shanhorey NA, Abdel-Hamid AME, Ali HM, Elshikh MS. Salicylic acid-regulated antioxidant mechanisms and gene expression enhance rosemary performance under saline conditions. Frontiers in Physiology. 2017;**8**:716

[27] El-Esawi M, Arthaut L, Jourdan N, d'Harlingue A, Martino C, Ahmad M. Blue-light induced biosynthesis of ROS contributes to the signaling mechanism of *Arabidopsis* cryptochrome. Scientific Reports. 2017;**7**:13875

[28] El-Esawi MA, Elkelish A, Elansary HO, Ali HM, Elshikh M, Witczak I, et al. Genetic transformation and hairy root induction enhance the antioxidant potential of *Lactuca serriola* L. Oxidative Medicine and Cellular Longevity. 2017;**2017**. Article ID: 5604746, 8 pages

[29] El-Esawi MA, Germaine K, Bourke P, Malone R. Genetic diversity and population structure of *Brassica oleracea* germplasm in Ireland using SSR markers. Comptes Rendus Biologies. 2016;**339**:133-140

[30] El-Esawi MA, Germaine K, Bourke P, Malone R. AFLP analysis of genetic diversity and phylogenetic

relationships of *Brassica oleracea* in Ireland. Comptes Rendus Biologies. 2016;**339**:163-170

[31] El-Esawi M, Glascoe A, Engle D, Ritz T, Link J, Ahmad M. Cellular metabolites modulate *in vivo* signaling of *Arabidopsis* cryptochrome-1. Plant Signaling & Behavior. 2015;**10**:e1063758

[32] El-Esawi MA, Mustafa A, Badr S, Sammour R. Isozyme analysis of genetic variability and population structure of *Lactuca* L. germplasm. Biochemical Systematics and Ecology. 2017;**70**:73-79

[33] El-Esawi MA, Al-Ghamdi AA, Ali HM, Alayafi AA. *Azospirillum lipoferum* FK1 confers improved salt tolerance in chickpea (*Cicer arietinum* L.) by modulating osmolytes, antioxidant machinery and stress-related genes expression. Environmental and Experimental Botany. 2019;**159**:55-65

[34] Jourdan N, Martino C, El-Esawi M, Witczak J, Bouchet P-E, d'Harlingue A, et al. Blue light dependent ROS formation by *Arabidopsis* cryptochrome-2 may contribute towards its signaling role. Plant Signaling & Behavior. 2015;**10**:e1042647

[35] Vwioko E, Adinkwu O, El-Esawi MA. Comparative physiological, biochemical and genetic responses to prolonged waterlogging stress in okra and maize given exogenous ethylene priming. Frontiers in Physiology. 2017;**8**:632

[36] El-Esawi MA, Al-Ghamdi AA, Ali HM, Ahmad M. Overexpression of *AtWRKY30* transcription factor enhances heat and drought stress tolerance in wheat (*Triticum aestivum* L.). Genes. 2019;**10**(2):163

[37] El-Esawi MA, Alayafi AA. Overexpression of *StDREB2* transcription factor enhances drought stress tolerance in cotton (*Gossypium barbadense* L.). Genes. 2019;**10**:142

Chapter 2

Novel Therapeutic Uses of Legume Crops in Southern South America

Renée Hersilia Fortunato, Virginia Fuentes Baluzzi, Fernando De Diego, Rodrigo T. Biagioni and Alejandro Daniel Esquivel

Abstract

The Argentine flora comprises more than 10,000 species, and many of them have been recorded as having medicinal, antimicrobial, and nutraceutical uses in humans as well as veterinary uses. In this chapter, native species/populations from the north of Argentina have been identified, selected, and characterized using morphological, chemical, and molecular techniques. *Bauhinia forficata* subsp. *pruinosa* was found to have anti-inflammatory, antidiabetic, diuretic, and analgesic activity and *Senna spectabilis* var. *spectabilis* has been found to have antibacterial, antibiofilm, antifungal, and antioxidant properties. The characterization and conservation of the native germplasm will allow us to propose future protocols of adaptation and technological processes to improve the quality of life in the rural areas and sustainable growth. This process will be achieved through a future integral and rational use that contemplates the conservation of the wild populations and their habitat. Thus, new resources will be generated, and the native flora of the country will gain value, strengthening the regional and territorial development of the agricultural and agroindustrial system. In addition, the domestication practices oriented to an integral management of the crop without extraction of the biological resource from the natural habitat minimize the impact of ecosystem degradation by overexploitation associated with landscape fragmentation.

Keywords: native species, medicinal, germplasm selection, conservation, propagation

1. Introduction

Leguminosae (Fabaceae) with close to 770 genera and ca. 20,000 species is the third-largest angiosperm family after Asteraceae and Orchidaceae [1]. It has a global distribution and high ecological and economic importance. Along with Poaceae (the grass family), Leguminosae is the most important plant family in the production of food for humans and livestock, as well as in the production of industrial products. The total world exports of pulses (crops harvested for their dry seeds) have more than doubled between 1990 and 2012, expanding from 6.6 to 13.4 million tons, and in 2012, the value of pulse exports was estimated at US$ 9.5 billion (Food and Agriculture Organization [FAO]: http://www.fao.org/3/a-i5389e.pdf). The Food and Agriculture Organization of the United Nations General Assembly designated 2016 as the International Year of Pulses to promote awareness of their nutritional benefits, importance in food security, and sustainable agriculture, and mitigate biodiversity loss and climate change http://www.fao.org/pulses-2016/en/. Meanwhile, from now on, 10 February will mark World Pulses Day, keeping alive the positive momentum surrounding these healthy, nutritious, protein-rich, and nitrogen-fixing legumes after FAO's successful 2016 International Year of Pulses Campaign. Growing pulses contributes to sustainable crop production: http://www.fao.org/news/story/en/item/1175295/icode/ and the legumes crops providing highly nutritious sources of protein and micronutrients that can greatly benefit health and livelihoods. This family is also uniquely important as fodder and green manure in both temperate and tropical regions and is used for their wood, tannins, oils, and resins, in the manufacture of varnishes, paints, dyes, and medicines, and in the horticultural trade. It has cosmopolitan in distribution, representing important ecological constituents in almost all biomes across the globe and occurs in even the most extreme habitats [2, 3]. The biomes represent significant elements in terms of species diversity and abundance, in lowland wet tropical forests in Africa, South America, and Asia [4], and they dominate dry forests and savannas throughout the tropics [5] and also occur in Mediterranean, desert, and temperate regions, up to high latitudes and at high elevations [6].

This high species richness is reflected in great morphological and chemical diversity, from which multiple uses are derived [7], that is, alkaloids, proanthocyanidins, and flavonoids can be present; pterocarpans are found only in legumes and they have significant antimicrobial, anticancerous, antiinflammatory, and antimalarial report activities [8]. There are also many legumes containing toxic and indigestible substances, which may be removed through various processing methods.

In America, the cultivation of Leguminosae dates from prehistoric times [9] and it has a great ethnobotanical importance in medicinal uses [10, 11].

Medicinal evaluation has indicated the importance of the environment as an important factor in the selection of useful resources by human populations [12].

In South America, there are reports of the use of Leguminosae species with different therapeutic activities [13–17].

In the South of South America, Argentina conserves ex situ collections of germplasm banks in the National Institute of Agricultural Technology (INTA), provincial banks, national universities, and the National Research Council (CONICET). Most of the extra INTA collections hold native species to protect biodiversity from anthropogenic impact. Anthropogenic activity has accentuated the degradation of natural habitats, environmental changes, landscape fragmentation, pollution, the expansion of the agricultural frontier, and over-exploitation. The Argentine Flora Catalog (http://www2.darwin.edu.ar/Proyectos/FloraArgentina/fa.htm) has records of more than 10,000 species; many of them have been recorded as having medicinal, antimicrobial, and nutraceutical properties, as well as veterinary uses. The available information has been obtained mainly from ethnobotanical, chemical, and biochemical studies; however, up to now, the chemical compound to obtain phytopharmaceutical products has been recorded only in a few species, and the promising biotypes have not been selected or adapted as new crops; therefore, there is no material available to develop medicinal products that ensure therapeutic efficacy.

The project that initiated the bioprospecting of new medicinal bioactive compounds and agrochemical products in Argentina was conducted at the Institute of Biological Resources (IRB), Natural Resources Research Center (CIRN), INTA, between 1993 and 2003. The proposal was based on national and international legislation (Argentina Constitution: 1994, and Convention on Biological Diversity, 1993, ratified in 1994) and was implemented through the agreement between INTA and University of Arizona, USA (Bioactive Agents from Dryland Plants of Latin America, INTA-Argentina/University of Arizona-USA, Grant UO1 TW00316 National Institutes of Health (NIH), National Science Foundation (NSF), U.S. Agency for International Development (USAID): International Cooperative Biodiversity Group (ICBG), 1993–2003). That evaluation allowed us to obtain new information on chemical and biochemical compounds of native species used in folk medicine [18, 19]. However, no species genotype or ecotype was introduced as a phytopharmaceutical product [19].

Within the frame of the project, and based on morphological, chemical, and molecular techniques (including the development of expression libraries), native species/populations were identified, selected, and characterized, via management in the natural habitat (*in situ*) and introduction in cultivation (*ex situ*). In addition, with the aim of conserving the identified biotypes, preservation was undertaken in the Germplasm Bank IRB, CIRN-INTA *ex-situ*), with subsequent introduction to cultivation in the different ecoregions where they grow. The characterization and conservation of the native germplasm will allow us to propose technological processes for the improvement of the quality of life in the rural territories; it will also promote sustainable growth through a future integral and rational use that contemplates the conservation of the wild populations and their habitat. In 2010 at the IRB, CIRN-INTA, we started the development of species with medicinal potential, each of them of regional importance in the central and northern regions of Argentina. Based on the results obtained, innovative lines (new-generation domestication) will be included.

a. *Bauhinia forficata* Link. subsp. *pruinosa* (Vog.) Fortunato & Wunderlin "cow's hoof, cow's paw, Brazilian orchid tree, unha de boi or pata de vaca: Brazil".

b. *Senna spectabilis* (DC.) H.S. Irwin & Barneby var. *spectabilis* "spectacular cassia, mhomba, carnaval, calceolaria shower, cassia, yellow shower, and Pau-deovelha, São-joão, Parica: Brazil".

2. Propagation

Bauhinia forficata subsp. *pruinosa* (=BFP): native to southern and eastern Brazil, Paraguay, Uruguay, and Argentina. In Argentina, it is distributed in the northeastern region, and through its introduction as ornamental, it has been naturalized also in the central and northwest regions. It is a deciduous to semi-evergreen tree up to 8 m tall, with twisted ascending branches that drop at the ends, an often-leaning trunk, and large, bilobed, dark green leaves; flowers are white, 8–13 cm in diameter, solitary, or arranged in axillary clusters; they bloom almost all summer. Fruits are dehiscent legumes with several flattened, oval, bright, and blackish seeds. It is used as ornamental and the leaves and young stems are consumed in infusions mainly as antidiabetic or hypoglycemic agent [20–25]. Anti-inflammatory, diuretic, and analgesic activities have also been reported in this species [26, 27]. In Southern South America, 10 species of *Bauhinia* are popularly used mainly to regulate glucose and lipids metabolism, but also for digestive, kidney, and urinary disturbances, for example, [28–31].

BFP has potential for the treatment of Diabetes *mellitus*. Pharmacological studies performed on different plant extracts or purified flavonoids in normoglycemic and hyperglycemic models in vivo in general have confirmed the hypoglycemic activity, although some contradictory evidence has been found [26, 32].

It is important to note that the species is unclearly identified in most assays (it is published as *B. forficata* or *B. candicans*, present synonym of BFP); as a result, flavonoid profiles and some tested activities present differences that should be analyzed in depth [33, 34]. In southern South America, the leaves and young stems are used to prepare a very popular tea due to their effective properties in reducing blood sugar levels. Wild specimens are collected, which violates the safety, quality, and efficacy requirements by WHO Guidelines on Good Agricultural and Collection Practices (GACP) for Medicinal Plants (http://apps.who.int/medicinedocs/es/d/Js5527s/) and causes loss of biological resources and its habitat. Likewise, it is emphasized that herbal medicines have a variable concentration of active ingredients, which depends on population diversity, the phenological stage at the time of harvest, and the edaphoclimatic conditions in which they grow [35–37]. In our recent studies, one population analyzed from plant materials collected in the Botanical Garden, INTA, presented five glycoside derivatives of kaempferol and quercetin [38], whereas another population/specimens belonging to the same garden did not show the same bioactivity (unpublished data).

The strategy of our proposal to solve this problem is to introduce different types of propagation—asexual and sexual—to evaluate the bioactivity and select the biotypes of BFP that show the best expression of bioactive compounds.

Senna spectabilis var. *spectabilis* (=SSS): tree native to tropical America. In Argentina, it is distributed in the northwestern provinces of Salta and Jujuy, in the zones of the subtropical forest, between 400 and 700 m above sea level. It is also cultivated in the south and east of Africa as an ornamental tree (http://www.buenosaires.gob.ar/noticias/senna-spectabilis). It is a small, rounded deciduous tree, 7–10 m (max. 15) tall, and 30 cm in trunk diameter, with a spreading crown, compound leaves, yellow, fragrant flowers, and cylindrical or flattened pods ending in a short, narrow point, hard, not splitting open or slightly on one side, pendulous, more or less cylindrical or slightly compressed, blackish, divided internally by partitions in compartments, aromatic, with numerous compressed and brown seeds. It has various medicinal uses: the leaves are used to treat skin diseases and to prevent constipation and intestinal parasite infections, as well as for headaches and migraines [39], and the leaves, stem, and roots of the taxonomical variety *excelsa* from Brazil are used in depressant and anticonvulsant activities [14]. There are records of antimicrobial activities of leaf extract against *Candida albicans* [40] and *Fusarium graminearum* [41]. In addition, it was shown that extracts of leaves, flowers, stems, and fruits have significant activity against *Escherichia coli*, *Bacillus cereus*, *Bacillus subtilis*, *Staphylococcus aureus*, and *Pseudomonas aeruginosa* [42, 43]. In phytochemicals, [44] isolated the alkaloid "(−) cassine" from the leaves and found antiinflammatory activity. On the other hand, [45] found that the anthraquinone compound emodin possessed antimicrobial activity and anti-inflammatory and laxative effects. There are also several reports [46] of populations of *S. spectabilis* from Brazil, whose fruits and flowers have pharmacological effects (analgesic and anti-inflammatory) due to the presence of more than 20 piperine and piperidine alkaloid compounds. It has been also presented important compounds report in leishmanicidal activity [47] and the strongly inhibit cell proliferation of hepatocellular carcinoma cells in alkaloids derived from flowers (in vitro antiproliferative and cytotoxic potentials of alkaloid mixture); if they represent promise antitumor compounds against liver cancer and should be considered for further anticancer in vivo studies [48].

In Argentina, antifungal activities were detected in studies of methanolic extracts from leaves of SSS against *Fusarium graminearum*. In addition, our studies determined antifungal activity against *F. verticillioides*, the causal agent of maize ear rot in fruits and flowers, adding a new record of bioactivity to this plant species [49]. The chromatographic reaction observed suggests that biological activities could be related to the anthraquinone emodin compound, as reported by [50] in *Senna*. *Fusarium* species are mycotoxin producers infecting crops during their development in the field, as well as their products during storage. The losses in the crops associated with fungi are due not only to yield reduction and the alteration of the commercial and nutritional quality, but also to the production of mycotoxins, which pose a risk for human and animal health [51].

2.1 Long-term Ex situ conservation

The germplasm of the populations that showed promising biotypes is preserved in the germplasm bank (Genetic Resources Conservation at the National Institute of Agricultural Technology (INTA): gene bank/germplasm bank).

According to the records, the seeds of *Bauhinia* and *Senna* are orthodox and can be conserved in long-term banks for the future access of selected germplasm [52]. All seed samples were dried in a controlled environment at between 5 and 20°C and with a relative humidity of 10–25%. After drying, they were placed in an airtight container suitable for long-term storage. In the cases of humidity, 25% of the total, they were stored in nonhermetic containers (medium-term: active collections). The long-term samples (base collections) were stored at −18° ± 3°C and a relative humidity of 15 ± 3%. About 25% of the samples under medium-term conditions (active collections) were preserved at 5 and 10°C and with relative humidity of 15 ± 3%.

2.2 Plants multiplication

BFP and SSS are commonly known species that are cultivated as ornamental tropical plants; however, no selection of the promising biotype with medicinal properties has been attempted. As a consequence, there is a lack of improved cultivars, and there are many difficulties in processing and marketing new crops.

Mature fruits were collected from specimens cultivated in the Botanical Garden, INTA (BFP) in March 2017 and from plant populations growing at different altitudes of the dry land forest in Salta province, northwestern Argentina, in March–April 2017 and from specimens cultivated in the Botanical Garden, INTA (SSS) in April 2017. The fruits of both species were packed in paper bags and transported to the laboratory, where they were thoroughly cleaned and stored in paper bags until the start of the experiments. The seeds were stored in cloth bags under room conditions (20–23°C and 50–75% relative humidity). Most experiments were conducted when the seeds were less than 1 year old (**Figure 1**). Only fully developed, undamaged seeds previously conditioned and disinfected were used for germination experiments: approximately 1000 seeds were mechanically scarified using fine sandpaper (**Figure 2**). The germination test was carried out in May with seeds previously hydrated for 24/48 hours [53]. The experiments were conducted at a constant temperature of 20°C in the greenhouse. Seeds were planted in soil in large pots (6 cm diameter × 10 cm height) and multi-celled plastic plug-trays: cell length 40 mm, width 41 mm, and depth 50 mm. The substrate mixture used was 70% sifted soil, 15% sand, and 15% perlite (**Figure 3**). The seeds were covered with 5 mm of substrate (**Figure 4**).

2.3 Preliminary results of adaptation

Seeds with 2-mm long radicles were considered germinated. Seedling emergence and height were recorded after 45 days.

In BFP: mechanical scarification increased germination (from 0.4 plants/day in non-scarified seeds to 1.2 plants/day in treated seeds); after 28 days, the percentage of germination was 15% for nonscarified seeds to 86% for scarified seeds. Therefore, scarification favored seedling uniformity (**Figure 5**).

Figure 1.
Mature seeds.

Figure 2.
Seeds scarified and hydrated.

SSS: plants emerged after 3–9 days, and 65% of plants germinated; the minimum and maximum temperatures were controlled. No frost damage was recorded, with the absolute minimum being 13°C. Planting was performed at the end of autumn, in a greenhouse with an automatic convection heating system. The seedlings (60 days

Figure 3.
Substrate mixture.

Figure 4.
Multi-celled plastic plugs.

Figure 5.
Seedling.

Figure 6.
Senna spectabilis *transplanted in individual pots.*

Figure 7.
Bauhinia forficata *subsp.* pruinosa*: specimens planted.*

Figure 8.
Planting scheme.

Figure 9.
Bauhinia forficata *subsp.* pruinosa*: annual growth post-transplant.*

old) were transplanted into individual pots: **Figure 6** (8 cm diameter × 19 cm height). They were ready for planting out 4–5 months later.

BFP: to test the propagation and management of plants that were already initially adapted, in September 2018, 3-year-old specimens were collected from Los Robles Park, Moreno, Buenos Aires (34°40′12.2″S 58°51′21.8″W). Those specimens were planted in spring 2015 with seeds from the established mother plants (**Figure 7**).

At the time of transplanting, the plants did not have leaves, since in the province of Buenos Aires, the species behaves as semi-tropical, due to the low winter temperatures (average minimum temperature 8°C). The specimens were approximately 1.52 m tall and 0.65 cm in diameter. Planting scheme was 8 plants/plot, 2 m between plants and 2 m between lines (**Figure 9**), in a randomized complete block design using R (www.r-project.org).

At present, survival, annual shoot growth, and different phenological stages post-transplanting are being monitored (**Figures 8** and **9**).

3. Conclusion

In the adapted specimens, harvesting of the organs of interest will be carried out in the four seasons, coinciding with different crop phenological stages (vegetative, flowering, and fruiting), to measure fresh and dry weight with the aim of determining biomass yield (kg/ha). In the next growing season, the harvested specimens will be evaluated according to regrowth capacity to determine the new available biomass.

Likewise, BFP might be reproduced asexually by the vegetative propagation technique, with plant cutting management. Agamic multiplication will be carried out in ligneous and semi-woody nursery cuttings, using different concentrations of rooters with synthetic auxins and different types of substrate [54].

4. Final remarks

Next-generation domestication: in the age of big data, gene editing, and next generation sequencing (NGS), we have the opportunity to document the transition from wild plant to crop. NGS provides a powerful tool for discovery of domestication genes in crop plants and their wild relatives [55]. The accelerated domestication of new plant species as crops may be facilitated by this knowledge. Re-sequencing of domesticated genotypes can identify regions of low diversity associated with domestication. Novel allelic variation in close or distant relatives can be characterized by NGS; the results give support to the selection and adaptation as new crops and ensure that biomass with therapeutic efficacy will be obtained. The characterization and evaluation of these model species have a phase of construction and validation of genomic cDNA expression libraries with the objective to accelerate the domestication of forest trees in a changing world. This strategy has allowed us to propose plans and technological processes to improve the quality of life in rural territories and to support sustainable growth of wild plant populations and their ecosystem.

Acknowledgements

This work was supported by the National Institute of Agricultural Technology (INTA), the National Research Council (CONICET) and Faculty of Agronomy and Agri-Food Sciences, University of Moron. This evaluation was included in the Project PNHFA-1106094.

The author also thanks María Monsalvo for her adviser during the figure editions.

Author details

Renée Hersilia Fortunato[1]*, Virginia Fuentes Baluzzi[2], Fernando De Diego[3], Rodrigo T. Biagioni[4] and Alejandro Daniel Esquivel[5]

1 Faculty of Agronomy and Agri-Food Sciences, University of Moron, Institute of Biological Resources (IRB), Natural Resources Research Center (CIRN), National Institute of Agricultural Technology (INTA), National Research Council (CONICET), Hurlingham, Buenos Aires, Argentina

2 National Research Council (CONICET) and National Institute of Agricultural Technology (INTA), Moreno, Buenos Aires, Argentina

3 Research Council (CONICET) and National Institute of Agricultural Technology (INTA), Institute of Biological Resources (IRB), Hurlingham, Buenos Aires, Argentina

4 Faculty of Agronomy and Agri-Food Sciences, University of Morón, Moron, Argentina

5 Institute of Biological Resources (IRB), Natural Resources Research Center (CIRN), National Institute of Agricultural Technology (INTA), Hurlingham, Buenos Aires, Argentina

*Address all correspondence to: fortunato.renee@inta.gob.ar

References

[1] Judd WS, Campbell CS, Kellogg EA, Stevens PF, Donoghue MJ, editors. Plant Systematics: A Phylogenetic Approach. Sunderland, Mass, USA: Sinauer Associates; 2002. 576 p. ISBN: 0-87893-403-0

[2] Schrire BD, Lavin M, Lewis GP. Global distribution patterns of the Leguminosae: Insights from recent phylogenies. Biologiske Skrifter. 2005;**55**:375-422. ISBN: 8773043044

[3] Schrire BD, Lewis GP, Lavin M. Biogeography of the Leguminosae. In: Lewis GP, Schrire B, MacKinder B, Lock M, editors. Legumes of the World. Richmond, U.K.: Royal Botanic Gardens, Kew; 2005. pp. 21-54. DOI: 10.1017/S0960428606190198

[4] Yahara T, Javadi F, Onoda Y, de Queiroz LP, Faith D, Prado DE, et al. Global legume diversity assessment: Concepts, key indicators, and strategies. Taxon. 2013;**62**:249-266. DOI: 10.12705/622.12

[5] Banda RK, Delgado-Salinas A, Dexter KG, Linares-Palomino R, Oliveira-Filho A, Prado D, et al. Plant diversity patterns in neotropical dry forests and their conservation implications. Science. 2016;**353**(6306):1383-1387. DOI: 10.1126/science.aaf5080

[6] Rundel PW, Arroyo MTK, Cowling RM, Keeley JE, Lamont BB, Vargas P. Mediterranean biomes: Evolution of their vegetation, floras, and climate. Annual Review of Ecology, Evolution, and Systematics. 2016;**47**(1):383-407. DOI: 10.1146/annurev-ecolsys-121415-032330

[7] Wink M. Evolution of secondary metabolites in legumes (Fabaceae). South African Journal of Botany. 2013;**89**:164-175. DOI: 10.1016/j.sajb.2013.06.006

[8] Selvam C, Jordan BC, Prakash S, Mutisya D, Thilagavathi R. Pterocarpan scaffold: A natural lead molecule with diverse pharmacological properties. European Journal of Medicinal Chemistry. 2017;**128**(10):219-236. DOI: 10.1016/j.ejmech.2017.01.023

[9] Zárate S. The archaeological remains of Leucaena (Fabaceae) revised. Economic Botany. 2000;**54**(4):477-499. DOI: 10.1007/BF02866547

[10] Leonti M, Ramirez F, Sticher O, Heinrich M. Medicinal flora of the Popoluca, Mexico: A botanical systematical perspective. Economic Botany. 2003;**57**(2):218-228. DOI: 10.1663/0013-0001(2003)057[0218:MFOTPM]2.0.CO;2

[11] Molares S, Ladio A. Ethnobotanical review of the Mapuche medicinal flora: Use patterns on a regional scale. Journal of Ethnopharmacology. 2009;**122**(2):251-260. DOI: 10.1016/j.jep.2009.01.003

[12] De Albuquerque UP, De Lucena RFP. Can apparency affect the use of plants by local people in tropical forests? Interciencia. 2005;**30**(8):506-511. ISSN: 0378-1844

[13] Roth I, Lindorf H, editors. South American Medicinal Plants, Botany, Remedial Properties and General Uses. New York: Springer-Verlag Berlin Heidelberg; 2002. 492 p. ISBN: 978-3-642-07544-5

[14] De Oliveira Silva F, De Vasconcelos Silva GM, Feng D, Mendes de Freitas R. Evaluation of central nervous system effects of iso-6-cassine isolated from *Senna spectabilis* var. excelsa (Schrad) in mice. Fitoterapia. 2011;**82**(2):255-259. DOI: 10.1016/j.fitote.2010.10.007

[15] Furlan FH, Zanata C, Dos Santos Damasceno E, Pintar de Oliveira L,

Da Silva LA, Moleta Colodel E, et al. Toxic myopathy and acute hepatic necrosis in cattle caused by ingestion of Senna obtusifolia (sicklepod; coffee senna) in Brazil. Toxicon. 2014;**92**:24-30. DOI: 10.1016/j.toxicon.2014.09.007

[16] Ferreira Macêdo MJ, Alves Ribeiro D, De Oliveira Santos M, Gonçalves de Macêdo D, Gonçalves Ferreira Macedo J, Vilar de Almeida B, et al. Fabaceae medicinal flora with therapeutic potential in Savanna areas in the Chapada do Araripe, Northeastern Brazil. Revista Brasileira de Farmacognosia. 2018;**8**(6):738-750. DOI: 10.1016/j.bjp.2018.06.010

[17] Macía MJ, García E, Vidaurre PJ. An ethnobotanical survey of medicinal plants commercialized in the markets of La Paz and El alto, Bolivia. Journal of Ethnopharmacology. 2005;**97**(2): 337-350. DOI: 10.1016/j.jep.2004.11.022

[18] Timmermann BN, Wächter G, Valcic S, Hutchinson B, Casler C, Henzel J, et al. The Latin American ICBG: The first five years. Pharmaceutical Biology. 1999;**37**(Suppl):35-54. ISSN: 1388-0209

[19] Timmermann BN. Biodiversity prospecting in drylands of Latin America. In: Pasternak D, Schlissel A, editors. Combating Desertification with Plants. New York: Springer Science + Business Media; 2001. pp. 97-111. DOI: 10.1007/978-1-4615-1327-8

[20] Fortunato RH, Varela BG, Castro MA, Nores MJ. Leaf venation pattern to recognize austral south American medicinal species of "cow's hoof" (*Bauhinia* L., Fabaceae). Revista Brasileira de Farmacognosia. 2017;**27**(2):158-161. DOI: 10.1016/j.bjp.2016.10.007

[21] Alisson H, Domingos E, Lindolpho C. Plantas Medicinais: Patas-De-Vaca. USP: Ser. Produtor Rural; 2016. 60 p. ISSN: 1414-4530

[22] De Sousa E, Zanatta L, Seifriz I, Creczynski-Pasa T, Pizzolatti M, Silva F. Hypoglycemic effect and antioxidant potential of kaempferol-3,7-O-(alpha)-dirhamnoside from Bauhinia forficata leaves. Journal of Natural Products. 2004;**67**:829-832. DOI: 10.1021/np030513u

[23] Pisoschi A, Pop A. The role of antioxidants in the chemistry of oxidative stress: A review. European Journal of Medicinal Chemistry. 2015;**97**:55-74. DOI: 10.1016/j.ejmech.2015.04.040

[24] Iribarren A, Pomilio A. Components of Bauhinia candicans. Journal of Natural Products. 1983;**46**(5):752-753. DOI: 10.1021/np50029a028

[25] Fuentes O, Alarcón J. Bauhinia candicans stimulation of glucose uptake in isolated gastric glands of normal and diabetic rabbits. Fitoterapia. 2006;**77**(4):271-275. DOI: 10.1016/j.fitote.2006.03.006

[26] Da Silva K, Cechinel Filho V. Plantas do gênero Bauhinia: composição química e potencial farmacológico. Química Nova. 2002;**25**(3):449-454. DOI: 10.1590/S0100-40422002000300018

[27] Rondina RVD, Bandoni AL, Coussio JD. Especies medicinales argentinas con potencial actividad analgésica. Dominguezia. 2008;**24**(1):47-69. ISSN: 1669-6859

[28] Almeida CFCBR, Lima e Silva TC, Amorim ELC, Maia MBS, Albuquerque UP. Life strategy and chemical composition as predictors of the selection of medicinal plants from the caatinga (Northeast Brazil). Journal of Arid Environments. 2005;**62**(1):127-142. DOI: 10.1016/j.jaridenv.2004.09.020

[29] Almeida CFCBR, Amorim ELC, Albuquerque UP, Maia MBS. Medicinal plants popularly used in the Xingó region—A semi-arid location in

Northeastern Brazil. Journal of Ethnobiology and Ethnomedicine. 2006;**2**:15. DOI: 10.1186/1746-4269-2-15

[30] Amorozo MCM. Uso e diversidade de plantas medicinais em Santo Antonio do Leverger, M.T., Brasil. Acta Botânica Brasílica. 2002;**16**(2):189-203. DOI: 10.1590/S0102-33062002000200006

[31] Silva MAB, Melo LVL, Ribeiro RV, Souza JPM, Lima JCS, Martins DTO, et al. Levantamento etnobotânico de plantas utilizadas como anti–hiperlipidêmicas e anorexígenas pela população de Nova Xavantina–MT, Brasil. Revista Brasileira de Farmacognosia. 2010;**20**(4):549-562. DOI: 10.1590/S0102-695X2010000400014

[32] Marques G, Rolim L, Alves L, Silva C, Soares L, Rolim-Neto P. Estado da arte de Bauhinia forficata Link (Fabaceae) como alternativa terapêutica para o tratamento do diabetes melittus. Revista de Ciências Farmacêuticas Báscia e Aplicada. 2013;**34**(3):313-320. ISSN: 1808-4532

[33] Farag MA, Sakna ST, El-Fiky NM, Shabana MM, Wessjohann LA. Phytochemical, antioxidant and antidiabetic evaluation of eight bauhinia L. species from Egypt using UHPLC-PDA-qTOF-MS and chemometrics. Phytochemistry. 2015;**119**:41-50. DOI: 10.1016/j.phytochem.2015.09.004

[34] Ferreres F, Gil-Izquierdo A, Vinholes J, Silva ST, Valentão P, Andrade PB. Bauhinia forficata link authenticity using flavonoids profile: Relation with their biological properties. Food Chemistry. 2012;**134**(2):894-904. DOI: 10.1016/j.foodchem.2012.02.201

[35] Barboza G, Cantero JJ, Nuñez C, Pacciaroni A, Ariza EL. Medicinal plants: A general review and a phytochemical and ethnopharmacological screening of the native argentine Flora. Kurtziana. 2009;**34**(1-2):7-365. ISSN: 1852-5962

[36] OMS. Directrices de la OMS Sobre Buenas Prácticas Agrícolas y de Recolección (BPAR) de Plantas Medicinales. Ginebra: Organización Mundial de la Salud. [Internet] 2003. 80 p. Available from: http://apps.who.int/medicinedocs/es/d/Js5527s/#Js5527s [Accessed: August 2003]

[37] Volpato GT, Damasceno DC, Rudge MVC, Padovani CR, Calderon IDMP. Effect of Bauhinia forficata aqueous extract on the maternal-fetal outcome and oxidative stress biomarkers of streptozotocin-induced diabetic rats. Journal of Ethnopharmacology. 2008;**116**(1):131-137. DOI: 10.1016/j.jep.2007.11.013

[38] Santos M, Fortunato RH, Spotorno VG. Analysis of flavonoid glycosides with potential medicinal properties on Bauhinia uruguayensis and Bauhinia forficata subspecies pruinosa. Natural Product Research. 2018;**5**:1-5. DOI: 10.1080/14786419.2018.1460826

[39] Jothy SL, Torey A, Darah I, Choong YS, Saravanan D, Chen Y, et al. Cassia spectabilis (DC) Irwin et barn: A promising traditional herb in health improvement. Molecules. 2012;**17**:10292-10305. DOI: 10.3390/molecules170910292

[40] Torey A, Sasidharan S. Anti-candida albicans biofilm activity by Cassia spectabilis standardized methanol extract: An ultrastructural study. European Review for Medical and Pharmacological Sciences. 2011;**15**: 875-882. ISSN: 11283602

[41] Salvat AE. Actividad antifúngica de extractos de plantas autóctonas sobre Aspergillus parasiticus y *Fusarium graminearum* [thesis]. Buenos Aires, San Martín: Universidad Nacional de San Martín; 2010

[42] Sangethaa S, Sasidharanb S, Zurainia Z, Suryania S. Antioxidant activity of methanolic extracts of cassia surattensis. Pharmacology. 2008;**2**: 829-838. ISSN: 1827-8620

[43] Chukeatirote E, Hanpattanakit P, Kaprom A, Tovaranonte J. Antimicrobial activity of *Senna spectabilis* and *S. Tora*. Journal of Plant Science. 2007;**2**(1): 123-126. DOI: 10.3923/jps.2007.123.126

[44] Silva GH, Teles HL, Zanardi LM, Young MCM, Eberlin MN, Hadad R, et al. Cadinane sesquiterpenoids of Phomopsis cassiae, an endophytic fungus associated with Cassia spectabilis (Leguminosae). Phytochemistry. 2006;**67**:1964-1969. DOI: 10.1016/j.phytochem.2006.06.004

[45] Guarize L, Da Costa JC, Dutra LB, Mendes RF, Lima IVA, Scio E. Anti-inflammatory, laxative and intestinal motility effects of Senna macranthera leaves. Natural Product Research. 2012;**26**(4):331-343. DOI: 10.1080/14786411003754264

[46] Viegas Júnior C, Rezende A, Silva DHS, Castro-Gambôa I, Bolzani VS. Aspectos químicos, biológicos e etnofarmacológicos do gênero Cassia. Quimica Nova. 2006;**29**(6):1279-1286. DOI: 10.1590/S0100-40422006000600025

[47] De Albuquerque Melo GM, Rodrigues Silva MC, Pereira Guimarães T, Mantovani Pinheiro K, Barbosa Brito da Matta C, Cavalcanti de Queiroz A, et al. Leishmanicidal activity of the crude extract, fractions and major piperidine alkaloids from the flowers of *Senna spectabilis*. Phytomedicine. 2014;**21**(3):277-281. DOI: 10.1016/j.phymed.2013.09.024

[48] Pereira RM, Ferreira-Silva GÁ, Pivatto M, Santos Lde Á, Bolzani Vda S, Chagas de Paula DA, et al. Alkaloids derived from flowers of *Senna spectabilis*, (−)-cassine and (−)-spectaline, have antiproliferative activity on HepG2 cells for inducing cell cycle arrest in G1/S transition through ERK inactivation and downregulation of cyclin D1 expression. Toxicology In Vitro. 2016;**31**:86-92. DOI: 10.1016/j.tiv.2015.11.018

[49] Di Ciaccio LS, Fortunato RH, Salvat AE. Actividad antifúngica de especies del género Senna (Caesalpinoideae, Leguminosae) del norte de Argentina frente a Fusarium verticillioides. Revista de Investigaciones Agropecuarias. 2018;**44**(1):111-120. ISSN: 1669-2314

[50] Wagner H, Bladt S. Plant Drug Analysis. A Thin Layer Chromatography Atlas. 2nd ed. Munich, Germany: Springer-Verlag Berlin Heidelberg New York; 1996. 70 p. ISBN: 978-3-642-00574-9

[51] Food and Agriculture Organization of the United Nations. The State of Food and Agriculture, Clime Change, Agriculture and Food Security [Internet]. 2016. Available from: http://www.fao.org/3/a-i6030e.pdf [Accessed: 2016]. ISBN: 978-92-5-109374-0

[52] Organización de las Naciones Unidas para la Alimentación y la Agricultura. Comisión de Recursos Genéticos para la Alimentación y la Agricultura. Proyecto de Normas para Bancos de Germoplasma de Recursos Fitogenéticos para la Alimentación y la Agricultura, 14a Reunión Ordinaria [Internet]. 2013. p. 121. Available from: http://www.fao.org/docrep/meeting/027/mf804s.pdf [Accessed: April 19, 2013]

[53] Lopes JC, Gomes Barbosa L, Torres Capucho M. Germinação de sementes de Bauhinia spp. Floresta. 2007;**37**(2):1-10. DOI: 10.5380/rf.v37i2.8658

[54] Landis TD, Tinus RW, Barnett JP, Nisley RG. Propagación de Plantas. Manual de viveros para la producción de especies forestales en contenedores. In: Hernández J, Provenzal P, editors.

Manual Agrícola 674. Vol. 6. Servicio Forestal del Departamento de Agricultura de los Estados Unidos (USDA-FS); 1998. p. 206. Available from: https://docplayer.es/42816419-Manual-de-viveros-para-la-produccion-de-especies-forestales-en-contenedor-volumen-seis.html

[55] Van Tassel D, Kenneth A, Albrecht JD, Bever AA, Boe YB, Crews T, et al. Accelerating silphium domestication: An opportunity to develop new crop ideotypes and breeding strategies informed by multiple disciplines. Crop Science. 2017;**57**:1274-1284. DOI: 10.2135/cropsci2016.10.0834

Chapter 3

Ethnomedicinal Values of Legume Plants in Pakistan

Faisal Hussain and Farzana Usman

Abstract

The data on medicinal plants in the vegetation of Pakistan was studied and surveyed from September to November, 2018. Different ethnomedicinal species were recorded which are used by local inhabitants as a medicine, fodder, fuel, and for agricultural purpose. Many of the medicinal plants recorded are used for the treatment of two or more diseases by the local people. The family Fabaceae was dominant with respect to medicinal plants. The precious knowledge of medicinal flora is rapidly vanishing due to the illiteracy among the local people and also due to destruction of the medicinal plants. The present study was designed to convey the knowledge and importance of medicinal flora as well as traditional uses of such plants in daily life.

Keywords: legume, medicinal values, traditional uses, treatments

1. Introduction

The wide variation in geography, altitude, soil, climate, and culture have created a rich floristic diversity, and it is estimated that there are about 6000 species of higher plants in Pakistan [1]. Although, the country has about 6,000 species of wild plants of which about 400–600 species are considered to be medicinally important [2]. There are a number of papers and investigations which have been documented on ethnomedicinal and folk uses of flora in almost all local inhabitants of Pakistan [3–15]. Pakistan is considered as a promising agricultural country of Asia. It has fertile land often covered with dense vegetation. There are several medicinal species recorded from almost all parts of the country. Local people are using commonly available plants for the treatment of many diseases and maintenance of their health. However, introduction of allopathic and homeopathic drugs has decreased human dependency on medicinal plants for their folk uses [16].

2. Geography of the area

The Pakistan is geographically diverse and extremely temperate region with wide variety of wildlife. It covers an area of 881,913 km^2 (340,509 sq miles). The Pakistan is categorized into three geographic areas: the northern highlands, the Indus River Plain, and the Balochistan Plateau. The climate is diverse and varies from tropical to temperate, coastal south and with arid conditions. Sometime monsoon season with heavy rainfall and flooding, and during dry season significant less

Family	Genus	Species	Life-form	Folk name
Fabaceae	*Acacia* Lam.	*Acacia nilotica* (Lam.) Willd.	Phanerophyte	Bubar
	Albizia Durazz.	*Albizia lebbeck* (Linn.) Benth.	Phanerophyte	Sareehan
	Alhagi Adans.	*Alhagi maurorum* Medik.	Chamaephyte	Kandiro
	Dalbergia L. f.	*Dalbergia sissoo* Roxb.	Phanerophyte	Talehi
	Mimosa L.	*Mimosa pudica* L.	Chamaephyte	Sharam Booti
	Prosopis L.	*Prosopis juliflora* (Sw.) DC.	Phanerophyte	Deevi
	Prosopis L.	*Prosopis cineraria* (L.) Druce	Phanerophyte	Kandi
	Tamarindus L.	*Tamarindus indica* (L.) Druce	Phanerophyte	Gidamari
	Trigonella L.	*Trigonella foenum-graecum* L.	Therophyte	Hurbo

Table 1.
List of common ethnomedicinal legume plants of Pakistan.

rainfall and it seems drought conditions in whole country. Pakistan has four distinct seasons. Due to the diversity of the landscape and climate in Pakistan, It is rich in a wide variety of trees, plants and ethnomedicinal flora. Due to the lack of knowledge and awareness ethnomedicinal flora of Pakistan suffer from a number of problems and threats including highest rate of deforestation, pollution and adverse effects of the ecosystem.

The medicinal flora and vegetation are usually available in almost all rural and remote areas of Pakistan. Due to the poor administration policies, the medicinal flora is rapidly vanishing due to the reasons of overgrazing, degradation and devastation. The need of this study is crucial due to the record of folk uses/treatment against disease. It is alarming need to save this medicinal flora from under threatened factors and this flora should be cultivated and used in pharmacy and drugs industry for human health [17].

The collection of medicinal plants was undertaken during September to November-2018. Family, common name of plants, folk name of plant, habit, parts used, medicinal uses and traditional uses were documented through the interviews of local "Hakeems" (doctors) and experienced growers of field crops. All species of medicinal plants were identified in the Department of Botany, Federal Urdu University of Art, Science and Technology, Karachi and the voucher specimens have been deposited in the Botany Herbarium.

All species were identified and confirmed with the help of flora of Pakistan [18, 19].

During the present study, a total of nine species belonging to eight genera and one family of angiosperms were documented (**Table 1**). The data for habits and life-forms of plants including herb, shrub, climber and trees were recorded.

These species with their respective families, common name, folk name, parts used, medicinal and traditional uses are listed below and placed into crops, shrubs, herbs and trees.

2.1 *Trigonella foenum-graecum* L.

Family: Fabaceae
Common name: Fenugreek
Folk name: Hurbo
Habit: Annual plant
Parts used: Seeds and leaves

Medicinal uses: The leaves of fenugreek are used as vegetable and as well as salads. Whenever seeds of fenugreek are considered to warm the kidneys, disperse cold and alleviate pain.

Traditional uses: The seeds of fenugreek are swallowed early in the morning with hot water and used before brushing the teeth and eating something. It considered effective against therapeutic and healing joint pain.

2.2 *Alhagi maurorum* Medik.

Family: Fabaceae
Common name: Camelthorn-bush
Folk name: Kandiro
Habit: Shrub
Parts used: Whole plant
Medicinal uses: The plant of camel thorn acts as a blood purifier used in skin allergy and possesses antioxidant activity. It is also considered as a treatment for glandular tumors and has antiseptic properties.
Traditional uses: The extract of plants is used as a pain killer of bones.

2.3 *Mimosa pudica* L.

Family: Fabaceae
Common name: Touch me not
Folk name: Sharam booti
Habit: Herb
Parts used: Roots, leaves, and flowers
Medicinal uses: The roots of *M. pudica* are commonly used for the treatment of vaginal and uterine problems, fatigue, asthma, and blood diseases.
Traditional uses: The leaves and flowers are traditionally recommended for the prevention of fever, ulcer and piles.

2.4 *Acacia nilotica* (Lamk.) Willd.

Family: Fabaceae
Common name: Acacia
Folk name: Bubar, babul
Habit: Tree
Parts used: Bark, flower, leaves, gum, and fruit
Medicinal uses: The leaves and flowers are used against hepatitis, ulcer and infertility of women. The leaves and fruits provide control of diarrhea and dysentery.
Traditional uses: The young stem of acacia is used as a tooth stick for the remedies of toothache. Its bark is used for the control of cough.

2.5 *Albizia lebbeck* (Linn.) Benth.

Family: Fabaceae
Common name: Siris/rain tree
Folk name: Sareehan
Habit: Tree
Parts used: Leaves and seeds
Medicinal uses: The siris is used as antiasthmatic in tuberculosis and trauma.

Traditional uses: The leaves of siris are used for the treatment of eye infection, whereas seeds are effective against boils or pimples.

2.6 *Dalbergia sissoo* Roxb.

Family: Fabaceae
Common name: Rose wood, shesham
Folk name: Talehi
Habit: Tree
Parts used: Leaves
Medicinal uses: The leaves of rosewood are considered effective for the hotness of body.
Traditional uses: The fresh twigs of rosewood are traditionally applied to relieve the ringworm and foot pain.

2.7 *Prosopis cineraria* (L.) Druce

Family: Fabaceae
Common name: Khejri, Jandi, and Ghaf
Folk name: Kandi
Habit: Tree
Parts used: Fruits and leaves
Medicinal uses: Prosopis is medicinally used as an anthelmintic, tonic, leprosy and asthma.
Traditional uses: The paste of leaves is externally applied over the injuries or cuts. The smoking of dry leaves is fruitful for eyes pain.

2.8 *Prosopis juliflora* (Sw.) DC.

Family: Fabaceae
Common name: Velvet mesquite
Folk name: Deevi
Habit: Tree
Parts used: Fruits and leaves
Medicinal uses: The velvet mesquite is used as antibacterial agent in alcoholic extracts. It is used in the treatment of colds, diarrhea, flu, and head cold.
Traditional uses: The fruit is used for the treatment of measles, eye infection, sore throat and wounds.

2.9 *Tamarindus indica* (L.) Druce

Family: Fabaceae
Common name: Tamarind
Folk name: Gidamari
Habit: Tree
Parts used: Fruit and Leaves
Medicinal uses: It is used against the treatment of spermarrhea and hepatitis. It is also used as a natural coagulant.
Traditional uses: The fruit of tamarind is considered as a promoter to sex hormones in females.

The almost all above mentioned recorded plant species are major source of medicinal purpose and it is also included in the ingredients of drugs and pharmacy industry. Some chemical compounds and active ingredients are beneficial against several human diseases and pathogens including epidermal, stomata and respiratory organs. During present study, it is observed that local and remote areas people communities prefer to use the traditional medicines against the diseases and these are easily approachable for everyone. The extract of these herbal plants are cheaper and quite safe as compared to synthetic and antibiotic treatments. However, allopathic treatment is not affordable to everyone. Due to the lack of awareness, shortage of water, overgrazing of domestic animals, salinity, floods, low rainfall, illiteracy and poverty in country, the precious flora is under threatened.

3. Conclusion

Rapid of population growth is also a principal cause of diminishing the ethnomedicinal plant vegetation. Certain species such as *Prosopis cineraria, Calotropis procera*, and *Grewia asiatica* are disappearing day by day in Pakistan. The documentation and survey indicates that Pakistan has very high potential flora for ethnomedicinal purpose. Therefore it is an urgent need for our local communities and educated peoples that they should be directly involved in creating the awareness about medicinal plant vegetation and their significance.

Author details

Faisal Hussain* and Farzana Usman
Laboratory of Plant Pathology and Aerobiology, Department of Botany, Federal Urdu University of Art, Science and Technology, Karachi, Pakistan

*Address all correspondence to: faisalhussain@fuuast.edu.pk

References

[1] Nasir E, Ali SI. Flora of Pakistan (Fascicle Series). Pakistan: Department of Botany, University of Karachi; 1970–1989. 260p

[2] Hamayun M, Khan A, Khan MA. Common medicinal folk recipes of District Buner, NWFP, Pakistan. Ethnobotanical Leaflets. 2003;**31**:56-64

[3] Ahmad M, Qureshi R, Arshad M, Khan MA, Zafar M. Traditional herbal remedies used for the treatment of diabetes from district Attock (Pakistan). Pakistan Journal of Botany. 2009;**41**: 2777-2782

[4] Bhatti GR, Qureshi R, Shah M. Ethnobotany of Qadanwari of Nara Desert. Pakistan Journal of Botany. 2001;**33**(SI):801-812

[5] Goodman SM, Ghafoor A. The ethnobotany of Southern Balochistan, Pakistan with particular reference to medicinal plants. Fieldiana. 1992;**31**:1-84

[6] Hocking GM. Pakistan Medicinal Plants I. Qualitas Plantarum et Material Vegetabiles. 1958;**5**:145-153

[7] Hocking GM. Pakistan Medicinal Plants-IV. Qualitas Plantarum et Material Vegetabiles. 1962;**9**:103-119

[8] Leporatti ML, Lattanzi E. Traditional Phytotherapy on Coastal Area of Makran (Southern Pakistan). Fitoterapia. 1994;**65**:158-161

[9] Malik SM, Shan M, Marwat Q. Ecotaxonomical Evaluation of Valuable Plants of Balochistan, Pakistan. Islamabad: Pakistan Science Foundation; 1990. 280p. Project Rep. No. 123

[10] Shinwari ZK, Malik S. Plant wealth of Dera Bughti area. Progressive Farming. 1989;**9**:39-42

[11] Qureshi R, Bhatti GR. Ethnobotany of plants used by the Thari people of Nara Desert, Pakistan. Fitoterapia. 2008;**79**:468-473

[12] Qureshi R, Waheed A, Arshad M, Umbreen T. Medico-ethnobotany of Tehsil Chakwal. Pakistan Journal of Botany. 2009;**41**(2):529-538

[13] Qureshi R, Bhatti GR, Saeed A. Obnoxious-mankinds need. Hamdard Medicus. 2002;**XLV**(2):82-87

[14] Qureshi R, Bhatti GR, Memon RA. Ethnomedicinal uses of herbs from Nara Desert, Pakistan. Pakistan Journal of Botany. 2010;**42**:839-851

[15] Qureshi R, Maqsood M, Arshad M, Chaudhry AK. Ethnomedicinal uses of plants by the people of Kadhi areas of Khushab, Punjab, Pakistan. Pakistan Journal of Botany. 2011;**43**(1):121-133

[16] Bhanu KU, Rajadurai S, Nayudamma Y. Studies on the tannins of babul, *Acacia arabica*, bark. Australian Journal of Chemistry. 1998; **17**:803-809

[17] Memon AH, Rind FMA, Laghari MGH, Mughal UR, Memon N, Gilal RA, et al. Common folk medicinal and ethnomedicinal uses of thirty medicinal plants of Districts Dadu and Jamshoro, Sindh, Pakistan. Sindh University Research Journal (Science Series). 2008; **40**:89-108

[18] Nasir E, Ali SI. Flora of Pakistan. Islamabad: NARC; 1971–1995. 790p

[19] Ali SI, Qaiser M. Flora of Pakistan. No. 194-216. Karachi, Pakistan: University of Karachi; 1993-2009. 459 p

Chapter 4

Starch Granules from Cowpea, Black, and Carioca Beans in Raw and Cooked Forms

Joyce Aparecida Tavares de Miranda, Lucia Maria Jaeger de Carvalho, Izabela Miranda de Castro, José Luiz Viana de Carvalho, André Luiz de Alcântara Guimarães and Ana Cláudia de Macêdo Vieira

Abstract

Starch applications in food systems are mainly influenced by solubility, gelatinization, paste viscosity, digestibility, and retrogradation. These characteristics result from properties such as the size and shape of granules, amylose and amylopectin contents, distribution of polymer chains, degree of crystallinity, and extraction of waste. In beans, the percentage of starch contents on dry basis is between 45 and 60%, being 24–65% amylose. This chapter evaluated the structure of common beans starch granules (*Phaseolus vulgaris*) and cowpea (*Vigna unguiculata*) in raw and cooked forms, by optical microscopy (OM) and scanning electron microscopy (SEM). Thus it was possible to observe the gelatinization of the starch granules especially in cowpea and carioca beans, as well as the "hard-to-cook" phenomenon in the black beans.

Keywords: common beans, cowpea, starches, SEM, OM, X-ray diffraction

1. Introduction

Several studies published in recent years have shown that the source of starch can substantially influence various technological processes in the food industry, such as texture and water retention of food, as well as in vital metabolic processes of human nutrition, such as the glycemic response to food intake [1–3]. In addition, starch can also be applied industrially in the production of nanofilms and biodegradable plastics [4]. Therefore, rather than a simple energetic component, starch must be studied based on its chemical differentiation in order to direct and optimize their technological and nutritional application [1]. Starch is the major reserve substance of the Plantae kingdom and is synthesized by plastid organelles. In dry cereals grains are found 40–90% of dry weight, in legumes 30–50% of dry weight, in tubers 65–85% of dry weight, and immature or green fruits 40–70% of dry weight.

IntechOpen

As in the other species of starchy legumes, the dry starch content in the various bean cultivars is between 45 and 60%. The granules contain ellipsoidal or spherical forms, with varying sizes, and contain high amounts of amylose (24–65%) [5].

Starches differ from the others carbohydrates due to granule form, built by the polymerization and by dehydration of amylose and amylopectin polysaccharides [6, 7]. Applications of starch in food systems are mainly governed by their properties of solubility, gelatinization, paste viscosity, retrogradation, and digestibility. These properties, in turn, result from characteristics such as the size and shape of the granules, amylose and amylopectin contents, the distribution of the polymer chains, degree of crystallinity of the granule, and the presence of extraction residues. These characteristics may be closely related to the events associated with gelatinization and retrogradation, such as granule swelling; amylose and/or amylopectin leaching; loss of radial (birefringence), supramolecular (crystallinity), and molecular structure; and recrystallization [1]. The size and shape of the grains vary between species, and for determining the size of the granules, microscopic methods have been applied [8].

The starch granule when observed microscopically under polarized light presents a typical "Malta Cross" model, resulting from the birefringence of its crystalline regions, which characterizes the radial orientation of the macromolecules. The center of the cross, called hilum, is considered the original growth point of the granule. The granule material is present in the form of concentric growth rings, which are arranged in alternating in the crystalline and amorphous regions. The fusion of these crystals and the inclusion of water cause rupture of this organized structure, characterizing the gelatinization of the starch granules [1]. The starch granules can be classified as simple when each plastid contains a granule or compounds when many granules are inside each amyloplast, as in the case of legumes [6, 7]. The shape can be spherical, oval, and polyhedral; the size is between 2 and 100 mm; and the particle size distribution is classified as unimodal, bimodal, or trimodal, being characteristic of the botanical origin. The surface is flat and relatively impermeable to large molecules such as amylases, due to the packaging of the amylopectin chains. By SEM, it is possible to observe the presence of some striations and fissures. Porosity to water and small soluble molecules occurs due to the reversible expansion of the amorphous regions, which may penetrate the entire granule during hydration, to form a continuous gel phase. However, the presence of pores or channels allows the entry of hydrolytic enzymes and other large molecules into the granules [9, 10].

X-ray diffraction patterns demonstrate that native (unmodified) starch grains contain between 15 and 45% crystalline material, corresponding to two polyphorm (A or B) and one intermediate form (C); the classifications are based in water content variations and in the double-helix packaging configuration. X-ray diffraction technique allows to distinguish the three types of crystallinity for the granules which, depending on their shape and crystalline structure, are called A, B, and C. These patterns of crystallinity depend in part on the length of the amylopectin chains, the packing density within the granules, as well as the presence of water. Starches with type A crystallinity exhibit intensity peaks at 2θ diffraction angles at approximately 15.3, 17.1, 18.2, and 23.5°; type B at about 5.6, 14.4. 17.2, 22.2, and 24°; and type C at about 5.6, 15.3, 17.3, and 23.5°. There is also a fourth type of crystallinity, type V, formed by the crystallization of amylose with lipids, which shows peaks of intensity at the 2θ diffraction angles at approximately 12.6, 13.2, 19.4, and 20.6° [11].

Type A crystallinity is described as a highly condensed and crystalline monocyclic cell unit, wherein 12 glucose residues from 2 chains in the counterclockwise direction harbor 4 molecules of water between the helices. Type A structure has amylopectin of chain lengths of 23–29 glucose units. Hydrogen bonding between the hydroxyl groups of the amylopectin molecule is responsible for the formation of the outer helical structure, among which linear chains of amylose moieties are packed through hydrogen bonds with amylopectin outer chains. This polyphorma occurs in most cereals such as corn, rice, wheat, and oats. Type B crystallinity is more clearly defined as being composed of a basic unit of chains which are packaged in a hexagonal array, wherein the cellular unit has two double helices counterclockwise, aligned and arranged in parallel and has amylopectin chain lengths of 30–44 glucose molecules, containing 36 molecules of water for every 12 glucose residues, with half of that water being tightly bound to the double helices and the other half being concentrated in a screw shaft. In addition to being considered rich in amylose, these types of starch have similar shapes and sizes and are resistant to hydrolysis, both enzymatic and acidic. The C-type structure is an amylopectin-containing intermediate structure of chain lengths of 26–29 glucose units. It is common to some roots and legumes. Starches with type A crystallinity are more susceptible to hydrolysis due to the presence of surface pores permeable to certain enzymes, whereas type B grains have protective shells, called crystalline blocks. However, type B crystals have a lower melting temperature (77°C) when compared to crystals A, of 90°C [10–15].

Some structural features such as amylose content, length, and distribution of amylopectin chain and degree of crystallinity in the granule may be related to the events responsible for gelatinization and retrogradation, such as granule swelling, amylose or amylopectin leaching and loss of molecular structure, birefringence, crystallinity, and recrystallization (Rupollo, 2011). A lot of techniques have been used to evaluate the behavior of starch against gelatinization. Methods use employing polarized light microscopy, X-ray diffraction, small-angle neutron scattering, and differential scanning calorimetry (DSC) and the viscosity assessment of starch pastes using equipment such as Rapid Visco Analyzer (RVA) and Brabender Visco-Amylo-Graph [10, 12].

Figueroa et al. [16] carried out a study about the thermal property characterization of cowpea, carioca, and white and black beans using DSC. The analysis provides quantitative measurements of the heat flux associated with gelatinization, which is represented by means of endothermic peaks in a characteristic range for each botanical source [17]. According to the study, the enthalpies of gelatinization showed to be significantly different (p-ANOVA <0.05) for all the samples; notice that the bean starch had the largest enthalpy, while the black bean had the lowest, being this property inversely proportional to the stability of the granule. The authors also conducted a study about the characterization of starch paste properties of the same bean varieties using an RVA, and they observed that the samples differed from each other, except for the minimum viscosity, for the carioca bean starches. According to the authors, this study assisted in a better structural and behavioral characterization of the starch granules these four different bean varieties, and the obtained results were taken as important parameters in the process and development of new food or technological products, with different applications of beans and their derivatives (flours and starch).

As time goes on for storage and cooling, gelatinized starch molecules are losing energy, and hydrogen bonds become stronger. In this way, the chains begin to reassociate in a more ordered state. This reassociation culminates in the formation of simple and double helices, resulting in junction zones between the molecules, forming crystalline areas. This phenomenon is called retrogradation and is influenced by factors such as temperature and storage time, pH, starch source, and presence of other components (lipids, electrolytes, sugars) and processing conditions. The amylose that had been exuded from the swollen granules forms a network by association with chains surrounding the gelatinized granules. As a consequence, the paste viscosity increases by converting to a cloudy viscoelastic system, or to an opaque elastic gel, precipitation of insoluble starch crystals leading to phase separation. The main influence of retrogradation is observed in the texture and digestibility of food containing starch, such as bakery products, and in the loss of water (syneresis) of some desserts that use starch as a thickener. As for digestibility, the retrogradation can be related to the lower availability of nutrients to the digestive enzymes, resulting in a lower glycemic response [1].

In a study involving the modification of starch concentration in cowpea by heat treatment, storage, and freezing, Salgado et al. [18] observed that the type of cooking, degree of maturation, and storage time exerted visible effects in the production of this starch fraction. Thermal processing altered the original morphological appearance and crystallinity pattern. According to the authors, the phenomenon of gelatinization, followed by retrogradation, can be considered beneficial from the nutritional point of view because they increase the dietary fiber content. This fact confirms a functional property attributed to the food, especially considering the role played by the short-chain fatty acids produced during the fermentation of resistant starch by bacteria present in the human colon.

Legume pulses stored under high temperature and high humidity adverse conditions can develop an increasing cooking time, characterized by prolonging cotyledon softening, phenomenon known as hard to cook (HTC). Beans that have undergone this HTC effect require increasing the cost of energy (fuel) for the preparation and are less acceptable to the consumer due to changes in flavor, color, and texture, with decreased nutritional quality. Several hypotheses have been proposed to explain the cause of bean hardening; among them are oxidation or polymerization of lipids, formation of insoluble pectates, lignification of intermediate lamellae, and multiple mechanisms. Maurer et al. [19] studied fractions extracted from common black and red beans using Fourier transform infrared spectroscopy (FT-IR). The samples were stored under three conditions: control at 4°C, HTC at 29°C and 65% humidity for 3.5 months, and HTC-chilled at 2°C and 79%. Two isolated fractions of the beans, the soluble pectin fraction and the insoluble residue of the cell wall, were analyzed. The immersion water and the cooking water of the beans were also studied. The results showed that in general, phenolic compounds were more associated with the fraction of soluble pectin of HTC beans than in the control bean. The results also showed that HTC-chilled beans contained higher concentrations of phenolic compounds than the control beans. Regarding immersion water, the authors observed that HTC-chilled beans and HTC had higher concentrations of absorbent compounds than control beans, indicating that they lost more constituents to the water. In addition, the results indicated that the mechanisms of reversibility of the HTC defect may be different from those involved in the development of the phenomenon.

Oliveira et al. [20] evaluated the cooking quality and nutritional composition of black and red beans, with and without storage under refrigeration. The grains were evaluated immediately after harvest, and after 6 months of storage, they were previously dried in a greenhouse (65–70°C) to a mean humidity of 13%. They were then packed in polyethylene bags and stored in a cold room at 0°C and 50% relative humidity. They evaluated the quality for cooking, the coloring of the seed coat, and the nutritional quality (protein, potassium, iron, and zinc). The authors observed that the evaluated beans maintained the quality for cooking and nutritional quality after 6 months of storage under refrigeration and that the clarity of the carioca bean tegument was altered.

Kruger et al. [21] evaluated the role that minerals play in the development of the HTC phenomenon in cowpea and its effect on in vitro bioaccessibility. The mineral distribution in normal and HTC cotyledons of cowpea bean was analyzed by proton-induced X-ray emission spectrometry (PIXE). The total phytate, tannin, and phenolic contents were analyzed together with the mineral bioaccessibility in vitro. The authors observed that in the HTC bean, Ca and Mg were more concentrated in the cell wall area of the medial lamella of the parenchyma cells. This, together with the reduction in phytate content, confirmed the phytase-phytate-mineral hypothesis as a mechanism for the development of the HTC phenomenon. Despite the reduction of phytate in stored cowpea, HTC decreased the bioaccessibility of Ca, Fe, and Zn in cowpea.

The goals of this work were to characterize the anatomical structure of common bean seeds (*Phaseolus vulgaris*) in the black and carioca varieties and cowpea (*Vigna unguiculata*), evaluate and compare the morphology of starch granules from this legumes by optical microscopy (OM) and SEM in raw and cooked forms, as well as to carry out the extraction of their starches and to characterize, by X-ray Diffraction, the models of crystalline structure.

2. Material and method

The beans were purchased from a retail food supply market in the city of Rio de Janeiro, RJ, and stored in a refrigerator with a temperature of approximately 4–8°C for about 12–18 months. The materials were subjected to the experiments, in the raw forms and cooked in a pressure cooker at 250°C for 25 minutes, and analyzed in duplicate.

For the characterization of the external morphology, samples of the abovementioned species and varieties were observed with the naked eye and with the aid of a magnifying glass 10X. Grains of both bean species were photographed using the Canon EOS Rebel T1i 15.0 megapixel camera. The images obtained were edited in Adobe® Photoshop® 7.0.1 software, and the boards were assembled using PowerPoint® 2007.

Anatomical studies on the raw materials were developed after hydration in a solution of 50% ethanol and glycerin (3.1), with the preparation of slides with transversal cuts in the freehand grains made in the middle region of the seeds. The cuttings were done with the help of blade stained with astra blue (C. not indicated) and safranin (C.I. 50,240) to prospect the structure of the organ under study. The slides were mounted in water. Subsequently, the samples were submitted to the paraffin infiltration technique, in order to obtain serial sections with the aid of a rotating microtome. The paraffined sections were affixed to the histological slides with Bissing adhesive and then dewaxed, dehydrated in an ethanolic series, and stained with astra blue (C. not indicated) and safranin (C.I. 50,240). After the ethanolic dehydration of the histological sections, the blades were assembled with synthetic resin (Entellan). Slides prepared from the included historesin material were stained with toluidine blue (C.I. 52,040), and the histological sections were adhered to the slides with water heated to 40°C. The images of the anatomical and histochemical analyses were obtained using a Quimis Q709ST-PLK microscope, with a capture system composed of a MOTICAM 2300 camera and Motic Images Plus® 2.0 software. Starch granules were observed using polarized light. The images obtained were edited in Adobe®Photoshop® software 7.0.1, and the boards were assembled using PowerPoint® 2007.

The cotyledon content of the raw and cooked bean samples was withdrawn for SEM studies. Prior to this procedure, the cooked beans were dehydrated at 40°C for 18 hours and then stored in a freezer at −10°C for 30 days and then were affixed in metal supports and coated with 30–35 nm of gold at 6.10^{-2} atm in gold sputter FL9496 Balzers metallizer. The observations and documentation of the material were carried out at the National Center of Structural Biology and Bioimaging (Centro Nacional de Biologia Estrutural e Bioimagem - CENABIO) of the Federal University of Rio de Janeiro (UFRJ), using Zeiss microscope, model EVO MA10, tungsten filament, working distance of 10 mm, and voltage of −15 Kv. The images were made with secondary electron detector.

Starch extraction was based on the method described by Wang and Wang (2004), with some modifications. The samples were ground in a laboratory mill (Perten, 3100) soaked in 0.1% NaOH solution in a ratio of 1:5 and allowed to stand for 20 hours. After dispersion, vigorous stirring in the blender was performed for 2 minutes. The resulting material was passed through a 63 μm sieve and centrifuged (Sorvall® RC 6 Plus centrifuge) at 1200 RPM for 5 minutes at room temperature (25°C ± 2). The supernatant was discarded, and the precipitate was resuspended in 0.1% NaOH solution and centrifuged again and the operation performed twice. The extracted starch was dispersed with distilled water and neutralized with 1 mol L^{-1} HCl to pH 6.5 and centrifuged. The sedimented material was resuspended in distilled water and centrifuged and the operation repeated twice. The resulting starch was oven-dried with air circulation at 40°C to 11% ± 0.5 humidity. The starch extraction yield was calculated on the difference between the dry flour masses before and after the starch isolation.

X-ray diffraction crystallinity tests were performed in the multiuser laboratory of the Chemistry Institute of the Federal University of Uberlândia (UFU). The diffractograms of the starches were obtained by an X-ray diffractometer (XRD-6000, Shimadzu, Brazil) in which the diffraction sweep region ranged from 5 to 30° with a target voltage of 30 kV and current of 30 mA. Scan speed was 1°min^{-1}. The relative crystallinity (RC) of the starch granules was calculated by the software XRD-6000 v. 5.2. The RC values of all bean samples were evaluated by GraphPad Prism® software, using the analysis of variance (ANOVA).

3. Results and discussion

3.1 Morphological characterization

Common bean and raw cowpea grains can be visualized in **Figure 1**, being (A) cowpea, (B) black beans, and (C) carioca bean.

Seed format characteristics and tegument coloration were within the range expected for species and varieties. Measurements of the materials studied revealed that the measurements were approximately 0.5–0.7 cm wide and 0.7–0.9 cm long for the three varieties of beans.

Figure 1.
Anatomical structure of bean pulse (A) cowpea, (B) black beans, and (C) carioca bean.

3.2 Optical microscopy (OM)

3.2.1 Cowpea

Figure 2 shows the optical microscopy of cowpea in cross section. Image A represents the integument of the raw grain prepared by manual cutting with an optical magnification of 50 x. Image B represents the tegument of the raw grain, prepared by the technique of infiltration in paraffin and serial cut with the help of a rotating microtome, both images with 50 x optical magnification. Image C refers to the cotyledon of the raw grain, and D refers to the cooked cotyledon under pressure; both images were observed using polarized light microscopy, with 10 x optical magnification, and slides prepared by freehand cut. Image E represents the cotyledon of the raw grain whose blade was prepared by freehand cut, and image F refers to cotyledon of the raw grain whose blade was prepared by serial cutting with the aid of a rotary microtome, preceded by paraffin infiltration; both images were obtained by optical microscopy with optical magnification of 50 x, using polarized light [22].

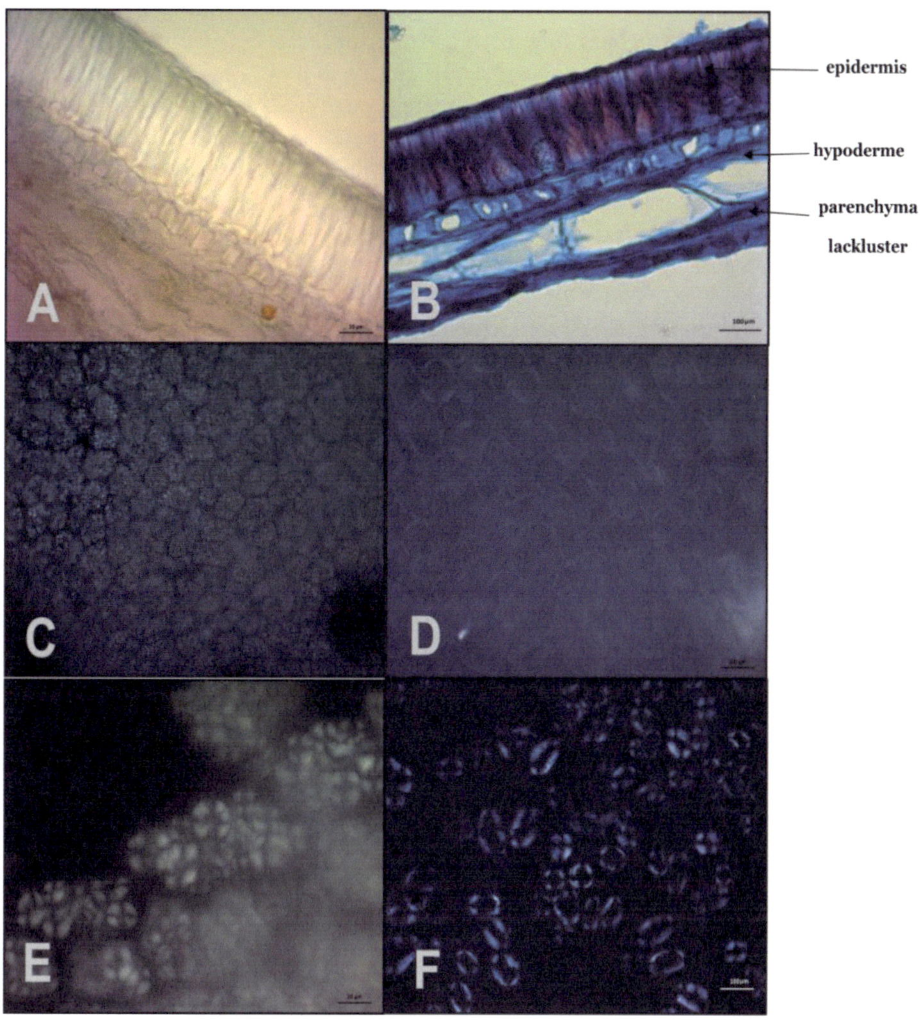

Figure 2.
Cross section of cowpea: (A) seed coat of the raw seed cut by free hand (50 x), (B) seed coat of the raw seed, cut with microtome (50 x), (C) cotyledon of raw seed (10 x), (D) cotyledon of cooked seed (10 x), (E) cotyledon of freehand cut seed (50 x), and (F) cotyledon of the raw seed cut with microtome (50 x).

In the integument images (A and B), three layers of tissues can be observed: epidermis, hypodermis, and lacunar parenchyma. The epidermis consists of two layers of flattened cells, with thickened walls. The hypodermis is composed of osteosclereids that are hourglass-shaped. The cells of the lacunar parenchyma have a shape close to the cylindrical, arranged with gaps between them. The thickness of the tegument is highly correlated with characters that reveal the size and shape of the seeds and water absorption capacity [23]. The images from SEM of raw whole cowpea seeds from Biaszczak et al. [24] revealed integument with an average thickness of about 90 μm, composed by palisade, glasshour and cells of lacunar parenchyma. In the tegument image whose material was previously infiltrated in paraffin (B), it is observed that the lacunar parenchyma presents rupture in the tissue, indicating that the presence of the reagents used in the embedding technique may have interfered in the sample structure, possibly due to the resumption of metabolic activities of seeds which broke their state of dormancy. Thus, the freehand cut allowed a better visualization of the structure, with the advantage of being a simple, efficient, and low-cost methodology, requiring no addition of organic solvents [23].

In the cotyledon images of the raw grain (C, E, and F) it is possible to perceive the great presence of amido within the amyloplasts, being coherent with the energy reserve function of this structure, since in beans as in other legumes, they constitute the main storage organs of the seed. The "Malta Cross" conformation of the starch resulting from the birefringence of the crystalline regions of the starch granule can be observed, as well as the characteristic spherical structure of the granule to that legume. Biaszczak et al. [24] reported that the cotyledon cells were rounded or elongated in the longitudinal axis, with an average size of 80 μm and had bimodal, elongated starch grains, firmly covered with protein material.

In the cooked grain image (D) the presence of starch grains exhibiting the characteristic "Malta Cross" is not observed, which evidences the loss of structural organization with the melting of the crystals. Such alteration is characteristic of the gelatinization process that occurs when the starch is submitted to temperatures higher than 50°C. Souza and Andrade [25] reported that after submission to temperatures above 75°C, there is no birefringence of starch grains of corn by optical microscopy on polarized light indicating loss of previously existing molecular ordering.

In the cotyledon image whose material was infiltrated in paraffin (E), the sharpness of the morphology of the starch granules is smaller than the image whose blade was prepared by free hand cut (F), suggesting once again that the technique manual cutting is more advantageous for bean seed samples.

3.2.2 Black bean

In **Figure 3**, optical microscopy of black beans in cross section can be observed. Image A represents the integument of the raw grain prepared by manual cutting with an optical magnification of 50 x. Image B represents the tegument of the raw grain, prepared by the technique of infiltration in paraffin and serial cut with the help of a rotating microtome, both images with 50 x optical magnification. Image C refers to the cotyledon of the raw grain, and D refers to the cooked cotyledon under pressure; both images were observed using polarized light microscopy, with 10 x optical magnification, and slides prepared by freehand cut. Image E represents the cotyledon of the raw grain, the blade of which was prepared by freehand cutting, and image F refers to the cotyledon of the raw grain, the blade of which was prepared by serial cutting with the aid of a rotary microtome; preceded by paraffin

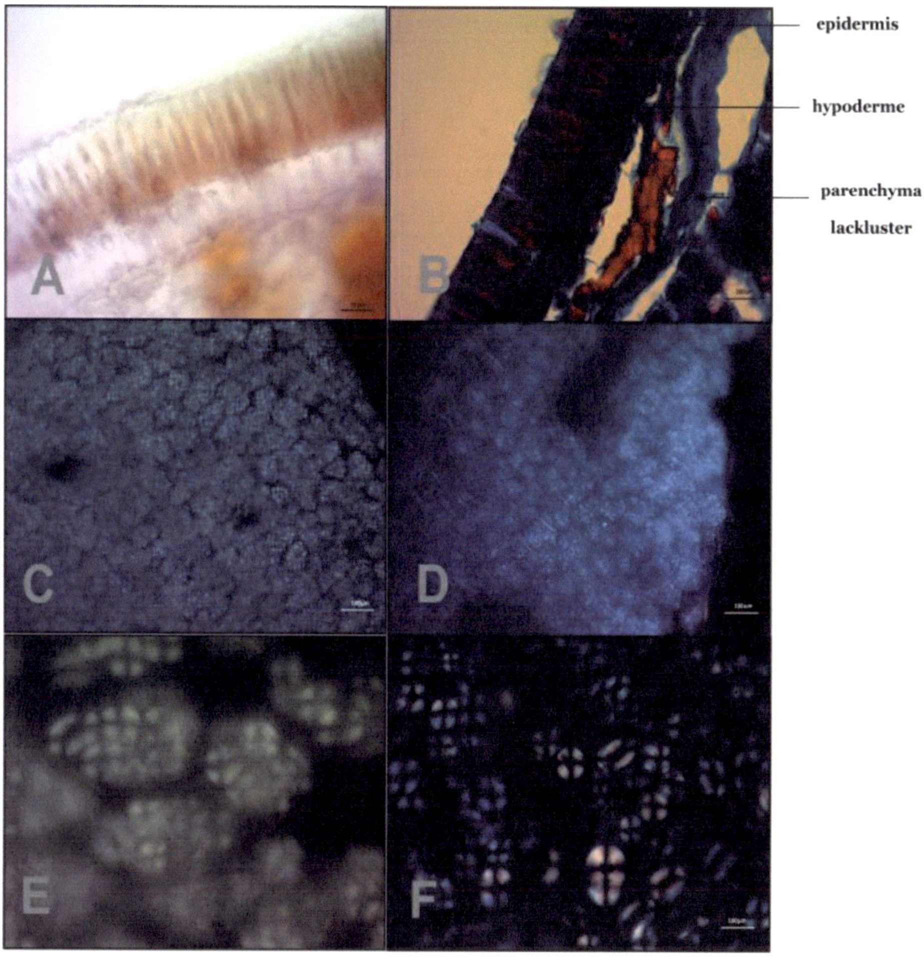

Figure 3.
Cross section of black beans: (A) seed coat of the raw seed, cut by free hand (50 x); (B) seed coat of the raw seed, cut with the help of a microtome (50 x); (C) cotyledon of raw seed (10 x); (D) cotyledon of boiled seed (10 x); (E) cotyledon of freehand cut seed (50 x); and (F) cotyledon of the raw seed, cut with the help of a microtome (50 x).

infiltration the images were obtained by optical microscopy with 50 x optical magnification using polarized light [22].

In the same way as in cowpea bean microscopy (**Figure 2**), three layers of tissues can be observed in the integument images (A and B): epidermis, hypodermis, and lacunar parenchyma. However, in the integument image whose material was previously infiltrated in paraffin (B), the lacunar parenchyma shows rupture in the tissue.

In the cotyledon images of the raw grain (C, E, and F), there is also a great presence of amido in the interior of the amyloplasts, its conformation of "Malta Cross," and the spherical structure of the granule. In the study by Ambigaipalan et al. [26], all black and carioca bean starch grains exhibited a strong birefringence pattern under polarized light, indicating that amylopectin crystallites are arranged radially within the bead at right angles to their interface with single reducing end group for the yarn. Weaker patterns would be indicative of double amylopectin helices disorganized within the crystalline lamellae of these grains. Chigwedere et al. [27] investigated the relative contributions of cotyledons and seed coats toward hardening of common beans were and the rate-limiting process which controls bean softening during cooking was determined. The authors suggested that the rate-determining process in bean softening relates to cell wall/middle lamella changes influencing pectin solubilization.

The presence of starch granules inside the amyloplasts is still observed in the cooked cotyledon image, suggesting that the cooking conditions were suitable for gelatinization, possibly due to a long storage period of the seeds, which characterizes an HTC phenomenon [19, 21].

3.2.3 Carioca bean

Figure 4 shows the optical microscopy of carioca beans in cross section. Image A represents the integument of the raw grain prepared by manual cutting with an optical magnification of 50 x. Image B represents the tegument of the raw grain, prepared by the technique of infiltration in paraffin and serial cut with rotating microtome, both images with 50 x optical magnification. Image C refers to the cotyledon of the raw grain, and D refers to the cooked cotyledon under pressure; both images are observed using polarized light microscopy, with 10 x optical magnification, and slides prepared by freehand cut. Image E represents the cotyledon of the raw grain whose blade was prepared by freehand cut, and image F refers to cotyledon of the raw grain whose blade was prepared by serial cutting with the aid of a rotary microtome, preceded by paraffin infiltration; both images were obtained by optical microscopy with optical magnification of 50 x, using polarized light [22].

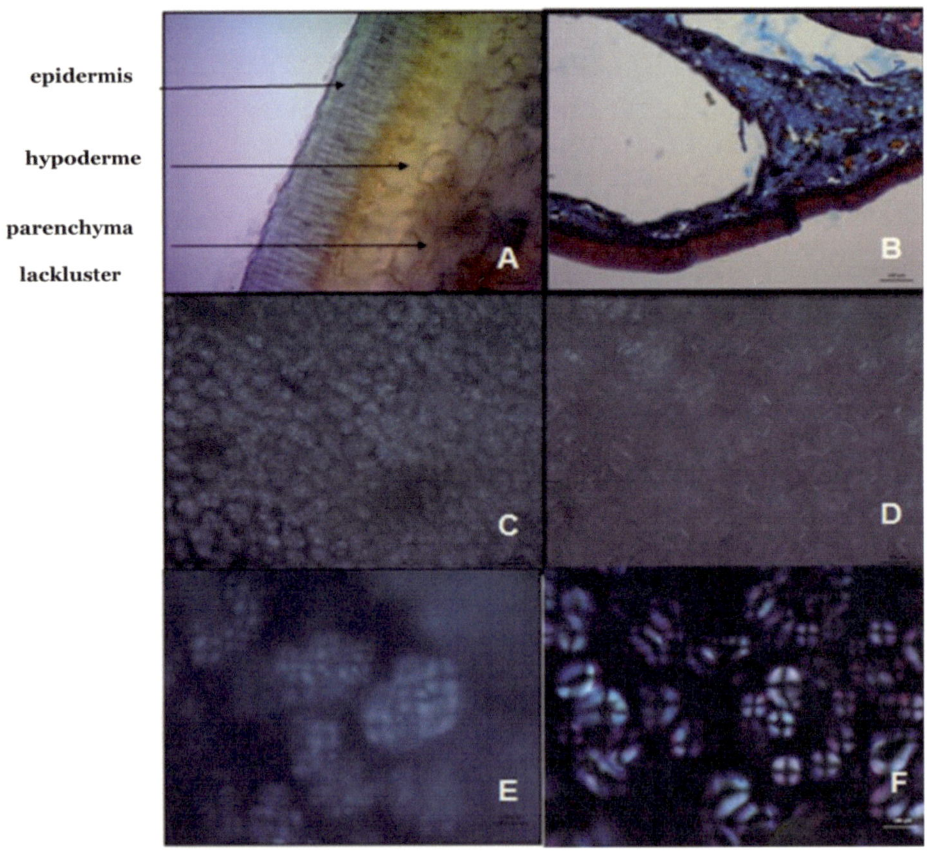

Figure 4.
Carioca bean cross section: (A) seed coat of the raw seed, cut by free hand (50 x); (B) seed coat of the raw seed, cut with microtome (50 x); (C) cotyledon of raw seed (10 x); (D) cotyledon of cocked seed (10 x); (E) cotyledon of freehand cut seed (50 x); and (F) cotyledon of the raw seed, cut with microtome (2.5 x).

Similar to the cowpea (**Figure 2**) and black bean microscopy (**Figure 3**), the layers of epidermis, hypodermis, and lacunar parenchyma can be observed in the tegument images (A and B). In the image of cotyledon of raw bean (C), a large amount of starch granules is also observed inside the amyloplasts, and in the image of cooked cotyledon (D), the absence of this structure is observed due to the phenomenon of gelatinization, also observed previously in the cowpea image (**Figure 2**). With 50 x optical magnification of cotyledon (images E and F), no notable differences were observed in the starch granules, between the blades that were prepared by freehand cutting or rotating microtome.

3.3 SEM

3.3.1 Cowpea

Figure 5 shows a cowpea endosperm SEM. In image A granules of starch attached to the cotyledon cell wall and its reniform shape (magnification 5.52 Kx) can be visualized, and in B bulges on the surface of the granule (magnification 30.36 Kx) can be perceived, both refer to raw samples. In the images of cooked samples with different optical amplifications (C 400 x and D 19.14 Kx), it is noticed that the starch granules were grouped, losing their crystalline structure due to gelatinization [28].

The morphological aspect of the starch observed in **Figure 5(A)** is in accordance with the description of Agunbiade and Longe [29], which confirmed that grain lengths in all samples were mostly larger than their widths. With the enlargement

Figure 5.
(A) Cowpea raw endosperm SEM (5.52 Kx), (B) cowpea raw endosperm SEM (30.36 Kx) granule surface, (C) cooked and gelatinized cowpea cotyledon (400 x), and (D) cooked cowpea cotyledon (19.14 Kx).

of the starch grains of the same **Figure 5(B)**, it is possible to perceive the presence of protrusions. The authors also compared the structure of cowpea, pigeon pea (*Cajanus cajan* L.) and yam bean (*Sphenostylis stenocarpa* L.), perceiving these grooves visible only in cowpea, being scarce in the yam bean, and almost imperceptible in pigeon pea. According to the authors, the morphological and legume starch characteristics are good indicators to identify their botanical origin and to detect if they are contaminated or adulterated with starches from other sources. They also observed that cowpea, pigeon pea, and yam bean exhibited appreciable shelf life stability, due to the low percentage of water and oil absorption.

Salgado et al. [30] observed that under the conditions in which their experiments were conducted, the morphological aspects of the starch grains were not influenced by the maturation stage of the grains. All presented a reniform shape, variable size between 11.8 μm and 26.7 μm, and smooth surface. Already the crystallinity pattern was higher in green beans than mature beans, as well as the percentage of resistant starch, whose test was based on the use of amylolytic enzymes.

3.3.2 Black bean

Figure 6 shows, by SEM, the black bean endosperm. In image A, the ellipsoid format of the starch granule is visualized (magnification 9.81 Kx), and in image B it is possible to observe the presence of cracks on the surface of the granule (magnification of 19.89 Kx), both refer to raw samples. The images of cooked samples (C magnification of 275 x and D and 29.01 Kx) reveal the stable structure of the black bean seed, suggesting a relation with the difficulties in its cooking (HTC phenomenon), requiring a longer cooking time for complete gelatinization, compared to the other samples studied [28].

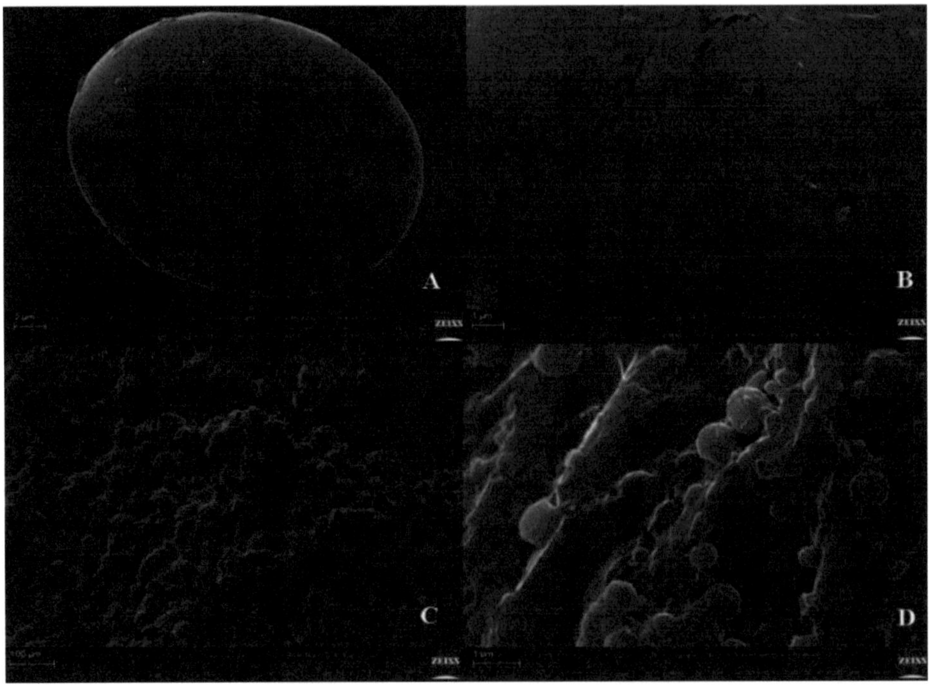

Figure 6.
(A) Black bean raw endosperm SEM (9.81 Kx), (B) black bean raw endosperm SEM (19.89 Kx) surface of granule, (C) cooked black bean cotyledon (275 x), and (D) unaccomplished gelatinization in cooked black bean cotyledon (29.01 Kx).

Ambigaipalan et al. [26], by SEM images, did not find the presence of cracks in black bean grains. Martínez-Preciado et al. [31] described the morphological structure of beans by SEM, observing that grains without the presence of fat had irregular oat-shaped starch granules with sizes of 10–40 μm in length and 10–25 μm in width, as well as small spherical beads of 10 μm. It was also observed that the starch grains were well defined and did not suffer any damage.

HTC phenomenon is one of the main obstacles to the consumption of beans grown in countries of Latin America and Africa, where ambient temperatures and relative humidity are high throughout the year, conditions that increase the possibility of occurrence of this phenomenon. At the microstructural level, the visible result of HTC seems to be related to the inability of the middle lamella of cotyledon cells to soften or dissolve and separate the cells [19, 21].

3.3.3 Carioca bean

In **Figure** 7, the SEM of the carioca bean endosperm is observed. In image A, the ellipsoid formed of the starch granules (magnification 6.00 Kx) is observed, and in image B one perceives protrusions on the surface of the granule (magnification 15.61 Kx), both refer to raw samples. In the images of cooked samples (C 370 x and D 15.61 Kx), with different optical amplifications, it is noticed that the starch granules were grouped, losing their crystalline structure due to gelatinization [28].

In the isolated starch granule images by SEM from Wang and Ratnayake [31] study, there was no evidence of starch damage, with no visible cracks or notches in the surfaces, and the presence of foreign materials was also not observed. All cultivars presented spherical, oval, or elliptic forms with smooth surfaces. According to the authors, generally, *P. vulgaris* starch granules have similar morphologies

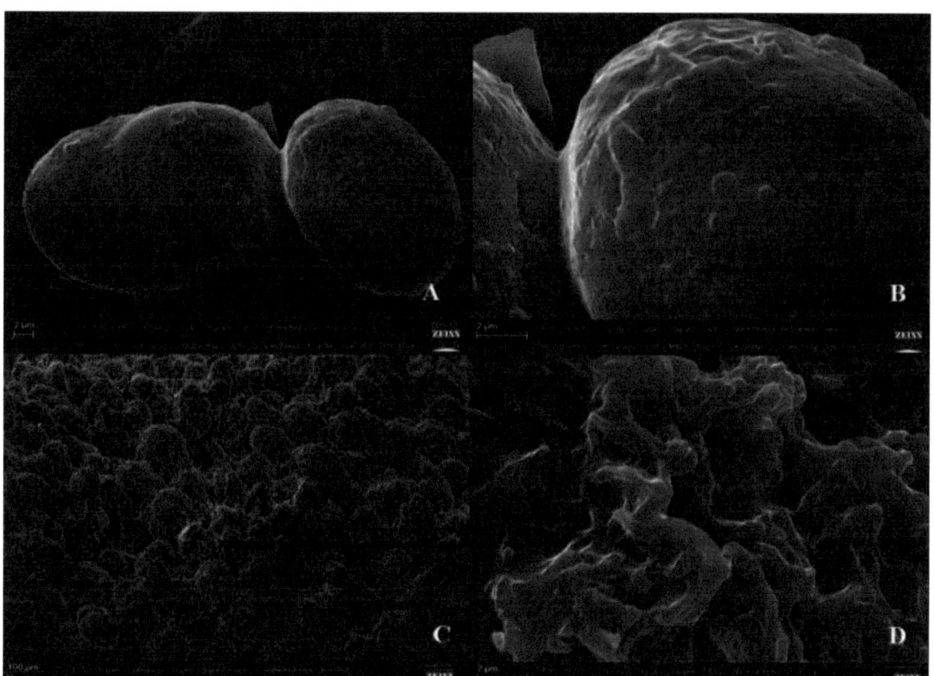

Figure 7.
(A) SEM of the raw carioca bean endosperm (6.00 Kx), (B) SEM of the endosperm of raw carioca bean (15.61 Kx), (C) cotyledon of cooked and gelatinized carioca bean (370 x), and (D) cotyledon of cooked carioca bean (15.61 Kx).

between their varieties but are very different from other starches such as tapioca and banana. The shape of the starch granules and size influence their functional properties, such as paste viscosity. A high viscosity is desirable for industrial uses, in which the purpose is the thickening function. Ambigaipalan et al. [26] did not find the presence of cracks in starch granules of carioca beans, as well as in black bean granules, both raw and evaluated by SEM.

Rupollo [17] analyzed by SEM the starch grains isolated from carioca beans stored for 360 days under three conditions: hermetically sealed at 5°C and atmosphere modified by nitrogen at 15°C and in a conventional atmosphere at 25°C. The author observed great similarity between the granules, even in different storage conditions of the seeds. However, the starch granules of the seeds stored in a conventional atmosphere at 25°C appeared to be more aggregate than the others. The influence of storage conditions on starch properties was verified through a joint data analysis, which is a multivariate technique used to evaluate how consumers develop preferences for products or services. The bean starch stored in a nitrogen-modified atmosphere at 15°C did not differ in solubility and gel properties compared to beans stored in a conventional atmosphere at 25°C. However, the gel properties of these two conditions differ from the hermetically packaged at 5°C, which presented lower crystallinity, as well as the swelling and heat power required for gelatinization. The grain starch stored in a nitrogen-modified atmosphere at 15°C, in turn, demonstrated lower crystallinity, swelling power, and heat required for gelatinization than grain stored in a conventional atmosphere at 25°C.

Vanier et al. [32] characterized starches from four common bean genotypes to use in production of biodegradable films. The authors observed that depending on the common bean genotype, a great variation on starch properties was found, which, in turn, clearly impacted on the characteristics of the starch-based films.

3.4 X-ray diffraction

The X-ray diffraction properties provide evidence of an ordered structure of the starch granule. The difference between crystallinities is associated with amylopectin, while amorphous regions are generally related to amylase. **Figure 8** presents the diffractograms of beans starches.

The starch isolated from black bean shows the highest peak values for the three evaluated. The diffraction angles 15, 17, and 23° represent the highest intensity peaks detected in the X-ray diffractograms, being even higher in 17° for all the analyzed starches in this work.

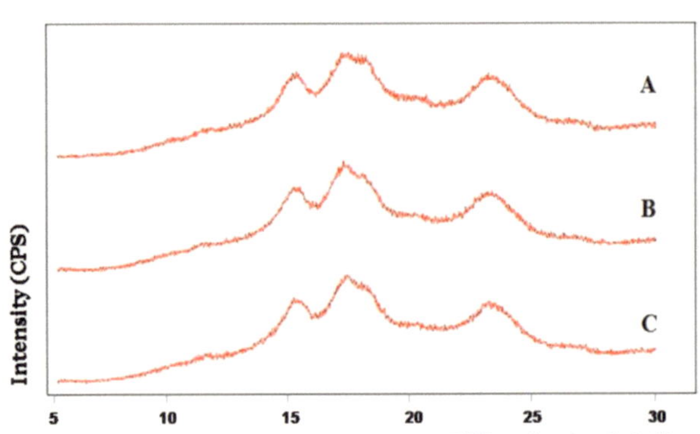

Figure 8.
Intensity of X-ray diffraction peaks of starch isolated from beans (A) cowpea, (B) black, and (C) carioca.

The yield of the starch isolation, the main peak intensities, and the relative crystallinity, verified in X-ray diffractograms, are shown in **Table 1**.

The relative crystallinity (RC) was in descending order, black bean (10.64%) > cowpea (10.57%) > carioca bean (10.50%), having varied significantly, considering the analysis of variance.

Gernat et al. [33] analyzed X-ray diffractograms of *Vicia faba* and *Pisum sativum* (bean and pea, respectively) compared to corn and potato starches, which are types A and B, respectively. The authors found, by means of a linear regression method, pea starch composed of 38.6% of type B and 61.4% of type A and 17.0% of bean starch of type B and 83.0% of type A. Garcia and Lajolo [34] analyzed the changes of HTC in starch grains and found a very strong birefringence in starch grains from HTC beans, suggesting that the starch isolated from these grains has a higher degree of crystallinity.

According to Hoover and Ratnayake [35], differences in relative crystallinity between starches are affected by crystal size, amount of crystalline regions (influenced by amylopectin chain content and length), and orientation of double helices in the crystalline domains and by degree interaction between double helices. In their work, all starches showed a pattern of type C X-rays, typical of legumes. The peak at $2\theta = 5.54$ (characteristic of type B starches) was more pronounced in pinto bean and black bean starches. Relative crystallinity followed the order: pinto beans > lentil > smooth pea > pea > black beans > white beans.

Type A pattern has the shortest amylopectin chain. Its structure is orthogonal and contains only eight molecules of water with few irregular connections, and amylose is distanced from amylopectin by an amorphous region, which is less dense and absorbs water more rapidly and is more susceptible to chemical and enzymatic modifications. In relation to the C pattern, a higher intensity of the diffractogram peak suggesting strong internal bonds of the molecules and a higher degree of association between the starch chains is observed [36].

Lawal and Adebowale [37] analyzed the physicochemical characteristics and thermal properties of chemically modified porcine bean (*Canavalia ensiformis*) and observed, in addition to the conventional type C, an increase in the intensity of starch diluted in acid solution. The authors did not observe significant differences between the X-ray pattern of native starch and modified derivatives.

Rupollo [17], evaluating the effects of storage conditions and time on the quality of carioca beans, observed that the starch of grains stored in a conventional atmosphere at 25°C were more influenced than the starch isolated from beans stored in modified atmosphere with nitrogen at 15°C, certainly due to the development of the HTC effect on grains stored in the conventional system.

Pinto [38] evaluated carioca bean starch submitted to different treatments and observed the following sequence regarding the degree of relative crystallinity

Bean	Starch yield (%)	Intensity (CPS*)			RC (%)
		15°	17°	23°	
Cowpea	16.04	1863	2214	1822	10.57
Black	16.85	1938	2358	1834	10.64
Carioca	30.24	1922	2306	1806	10.50

**CPS: counting per second.*

Table 1.
Starch yield, main peaks intensity, and relative crystallinity of isolated starches from cowpea, black, and carioca.

obtained with the different procedures: enzymatic hydrolysis > native starch > heating > ultrasound > low humidity heat treatment.

4. Conclusion

No OM differences were observed between the morphology of the starch grains of raw samples of cowpea, black beans, and carioca beans. The cotyledons of cooked carioca bean and cowpea samples completely loss the structural organization of the starch granules. In the cotyledons images of black common bean samples, cooked under the same conditions as the others (180°C in a pressure cooker for 45 minutes), the presence of the starch granule is still observed, suggesting the occurrence of the HTC, confirmed in this study by SEM, phenomenon whose should be of great relevance in the inspection of grains put on sale for consumption. With regard to the crystallinity studies of starch granules, by XRD, the diffraction angles found in this work are more consistent with the classification of the standard polyphorma A, and the RC was in descending order: black bean > cowpea > carioca bean.

Acknowledgements

The authors would like to thank Cappes for the financial support and Embrapa-CTAA, CENABIO-UFRJ, IQ-UFU, Laudo Laboratório Avícola for letting them use of the labs and equipment.

Author details

Joyce Aparecida Tavares de Miranda[1], Lucia Maria Jaeger de Carvalho[1*], Izabela Miranda de Castro[2], José Luiz Viana de Carvalho[2], André Luiz de Alcântara Guimarães[1] and Ana Cláudia de Macêdo Vieira[1]

1 Department of Natural Products and Food, Federal University of Rio de Janeiro (UFRJ), Rio de Janeiro, Brazil

2 Center for Technology and Food Analysis (CTAA), Laboratory of Residues and Contaminants, Brazilian Agricultural Research Corporation (EMBRAPA), Rio de Janeiro, Brazil

*Address all correspondence to: luciajaeger@gmail.com

References

[1] Denardin C, Silva L. Starch granules structure and its regards with physicochemical properties. Ciência Rural. 2009;**39**(3):945-954. DOI: 10.1590/S0103-84782009005000003

[2] Los FGB, Zielinski AAF, Wojeicchowski JP, Nogueira A, Demiate IM. Beans (*Phaseolus vulgaris* L.): Whole seeds with complex chemical composition. Current Opinion in Food Science. 2018;**19**:63-71. DOI: 10.1016/j.cofs.2018.01.010

[3] Rezende AA, Pacheco MTB, Silva VSND, Ferreira TAPDC. Nutritional and protein quality of dry Brazilian beans (*Phaseolus vulgaris* L.). Food Science and Technology. 2017;**38**(3):421-427. DOI: 10.1590/1678-457x.05917

[4] Schmidt V, Bertib F, Portob L, Laurindo J. Production of starch acetate films with addition of bacterial cellulose nanofibers. Chemical Engineering Transactions. 2013;**32**:2251-2256. DOI: 10.3303/1332376

[5] Shiga T, Lajolo F. Cell wall polysaccharides of common beans (*Phaseolus vulgaris* L.)—Composition and structure. Carbohydrate Polymers. 2006;**63**(1):1-12. DOI: 10.1016/j.carbpol.2005.06.025

[6] Lajolo F, Menezes E. Carbohidratos en alimentos regionales iberoamericanos. Revista Brasileira de Ciências Farmacêuticas. 2006;**42**(4):646. DOI: 10.1590/S1516-93322006000400018

[7] Eliasson A. Starch in Food—Structure, Function and Applications. New York: Boca Raton, CRC; 2004. p. 605. DOI: 10.1111/j.1745-4603.2005.00013.x

[8] Leonel M. Analysis of the shape and size of starch grains from different botanical species. Food Science and Technology. 2007;**27**(3):579-588. DOI: 10.1590/S0101-20612007000300024

[9] Tester R, Karkalas J, Qi X. Starch—Composition, fine structure and architecture. Journal of Cereal Science. 2004;**39**(2):151-165. DOI: 10.1016/j.jcs.2003.12.001

[10] Lima B, Cabral T, Neto R, Tavares M, Pierucci A. Estudo do amido de farinhas comerciais comestíveis. Polímeros. 2012;**22**(5):486-490. DOI: 10.1590/S0104-14282012005000062

[11] Sajilata M, Singhal R, Kulkarni P. Resistant starch–A review. Comprehensive Reviews in Food Science and Food Safety. 2006;**5**(1):1-17. DOI: 10.1111/j.1541-4337.2006.tb00076.x

[12] Sigh N, Singh J, Kaur L, Sodhi N, Gill B. Morphological, thermal and rheological properties of starches from different botanical sources. Food Chemistry. 2003;**81**(2):219-231. DOI: 10.1016/S0308-8146(02)00416-8

[13] Singh B, Singh JP, Shevkani K, Singh N, Kaur A. Bioactive constituents in pulses and their health benefits. Journal of Food Science and Technology. 2017;**54**:858-870. DOI: 10.1007/s13197-016-2391-9

[14] He W, Wei C. Progress in C-type starches from different plant sources. Food Hydrocolloids. 2017;**73**:162-175. DOI: 10.1016/j.foodhyd.2017.07.003

[15] Santiago-Ramos D, Figueroa-Cárdenas JD, Véles-Medina JJ, Salazar R. Physicochemical properties of nixtamalized black bean (*Phaseolus vulgaris* L.) flours. Food Chemistry. 2018;**240**:456-462. DOI: 10.1016/j.foodchem.2017.07.156

[16] Figueroa A, Lacerda L, Schnitzler E, Demiate I. Caracterização das propriedades de pasta e térmicas de

amidos de feijão Branco, Carioca, Fradinho e Preto. Brazilian Journal of Thermal Analysis. 2015;**4**(3):12-16. DOI: 10.18362/bjta.v4.i3.197

[17] Rupollo G. Efeitos das condições e do tempo de armazenamento na qualidade de grãos de feijão carioca [thesis]. Universidade Federal de Pelotas; 2011

[18] Salgado S, de Melo Filho A, Andrade S, Maciel G, Livera A, Guerra N. Modificação da concentração de amido resistente em feijão macassar (l. Walp) por tratamento hidrotérmico e congelamento. Food Science and Technology. 2005;**25**(2):259-264. DOI: 10.1590/S0101-20612005000200013

[19] Maurer G, Ozen B, Mauer L, Nielsen S. Analysis of hard-to-cook red and black common beans using Fourier transform infrared spectroscopy. Journal of Agricultural and Food Chemistry. 2004;**52**(6):1470-1477. DOI: 10.1021/jf035083s

[20] de Oliveira V, Ribeiro N, Maziero S, Cargnelutti Filho A, Jost E. Cooking quality and nutritional composition of common bean genotypes with and without stored refrigeration. Ciência Rural. 2011;**41**(5):746-752. DOI: 10.1590/S0103-84782011005000050

[21] Kruger J, Minnis-Ndimba R, Mtshali C, Minnaar A. Novel in situ evaluation of the role minerals play in the development of the hard-to-cook (HTC) defect of cowpeas and its effect on the in vitro mineral bioaccessibility. Food Chemistry. 2015;**174**:365-371. DOI: 10.1016/j.foodchem.2014.10.134

[22] Miranda JAT, Carvalho LMJ, Vieira ACM, Costa RPC, Guimarães ALA. Anatomical characterization and evaluation of starch granules in grain of black common bean and cowpea raw and cooked. Chemical Engineering Transactions. 2015;**44**:85-90. DOI: 10.3303/CET1544015

[23] Cavariani C, Toledo M, Rodella R, França-Neto J, Nakagawa J. Velocidade de hidratação em função de características do tegumento de sementes de soja de diferentes cutivares e localidades. Revista Brasileira de Sementes. 2009;**31**(1):30-39. DOI: 10.1590/S0101-31222009000100004

[24] Biaszczak W, Doblado R, Frías J, Vidal-Valverde C, Sadowska J, Fornal J. Microstructural and biochemical changes in raw and germinated cowpea seeds upon high-pressure treatment. Food Research International. 2007;**40**(3):415-423. DOI: 10.1016/j.foodres.2006.10.018

[25] Souza R, Andrade C. Investigação dos processos de gelatinização e extrusão de amido de milho. Polímeros: Ciência e Tecnologia. 2000;**10**(1):24-30. DOI: 10.1590/S0104-14282000000100006

[26] Ambigaipalan P, Hoover R, Donner E, Liu Q, Jaiswal S, Chibbar R, et al. Structure of faba bean, black bean and pinto bean starches at different levels of granule organization and their physicochemical properties. Food Research International. 2011;**44**(9):2962-2974. DOI: 10.1016/j.foodres.2011.07.006

[27] Chigwedere CM, Olaoye TF, Kyomugasho C, Jamsazzadeh Kerman Z, Pallares Pallares A, Van Loey AM, et al. Mechanistic insight into softening of Canadian wonder common beans (*Phaseolus vulgaris*) during cooking. Food Research International. 2018;**106**:522-531. DOI: 10.1016/j.foodres.2018.01.016

[28] Miranda JAT, Carvalho LMJ, Vieira ACM, Castro IM. Scanning electron microscopy and crystallinity of starches granules from cowpea, black and carioca beans in raw and cooked forms. Food Science and Technology. 2019. ID CTA-2018-0307.R1

[29] Vanier NL, Oliveira JP, Bruni GP, El Halal SLM, Villanova FA, Zavareze ER, et al. Characteristics of starch from different bean genotypes and its effect on biodegradable films. Journal of the Science of Food and Agriculture. 2018;**99**(3):1207-1214. DOI: 10.1002/jsfa.9292

[30] Agunbiade S, Longe O. The physico-functional characteristics of starches from cowpea (*Vigna unguiculata*), pigeon pea (*Cajanus cajan*) and yambean (*Sphenostylis stenocarpa*). Food Chemistry. 1999;**65**(4):469-474. DOI: 10.1016/S0308-8146(98)00200-3

[31] Martínez-Preciado A, Estrada-Girón Y, González-Álvarez A, Fernández V, Macías E, Soltero J. Physicochemical, morphological and rheological properties of canned bean pastes "negro Queretaro" variety (*Phaseolus vulgaris* L.). Journal of Food Science and Technology. 2012;**51**(9):1795-1805. DOI: 10.1007/s13197-012-0737-5

[32] Wang H, Ratnayake W. Physicochemical and thermal properties of *Phaseolus vulgaris* L. var.Great northern bean starch. Journal of Food Science. 2014;**79**(3):C295-C300. DOI: 10.1111/1750-3841.12357

[33] Gernat C, Radosta S, Damaschun G, Schierbaum F. Supramolecular structure of legume starches revealed by X-ray scattering. Starch-Stärke. 1990;**42**(5):175-178. DOI: 10.1002/star.19900420504

[34] Garcia E, Lajolo FM. Starch alterations in hard-to-cook beans (*Phaseolus vulgaris* L.). Journal of Food Chemistry. 1994;**42**:612-615. DOI: 10.1021/jf00039a002

[35] Hoover R, Ratnayake W. Starch characteristics of black bean, chick pea, lentil, navy bean and pinto bean cultivars grown in Canada. Food Chemistry. 2002;**78**(4):489-498. DOI: 10.1016/S0308-8146(02)00163-2

[36] Salgado S, Guerra N, Andrade S, Livera A. Caracterização físico-química do grânulo do amido do feijão-caupi. Food Science and Technology. 2005;**25**(3):525-530. DOI: 10.1590/S0101-20612005000300022

[37] Lawal O, Adebowale K. Physicochemical characteristics and thermal properties of chemically modified jack bean (*Canavalia ensiformis*) starch. Carbohydrate Polymers. 2005;**60**(3):331-341. DOI: 10.1016/j.carbpol.2005.01.011

[38] Pinto V. Pré-tratamentos na produção de nanocristais de amido de pinhão e feijão [thesis]. Universidade Federal de Pelotas; 2014

Chapter 5

Mungbean (*Vigna radiata* L. Wilczek): Retrospect and Prospects

Suhel Mehandi, Syed Mohd. Quatadah,
Sudhakar Prasad Mishra, Indra Prakash Singh, Nagmi Praveen and Namrata Dwivedi

Abstract

Mungbean (*Vigna radiata* L. Wilczek) is economically most important crop of *Vigna* group. It is also known as green gram, golden gram, moong, Chickasaw, Oregon pea, and chop suey bean and this legumes have a strategic position in Southeast Asian countries for nutritional security and sustainable crop production. Being rich in quality protein, minerals and vitamins, they are inseparable ingredients in the diets of a vast majority of Indian population. When supplemented with cereals, they provide a perfect mix of essential amino acids with high biological value. These crops have the ability to fix atmospheric nitrogen (58–109 kg per ha in kg per ha mungbean) in symbiotic association with Rhizobium bacteria, which enables them to meet their own nitrogen requirement and also benefit the succeeding crops. This crop has also been reported to smother weed flora appreciably (20–45%) when intercropped with tall cereals or pigeon-pea and consequently, minimize the cost incurred on weed control. On account of short duration and photo-thermo insensitivity, they are considered excellent crops for crop intensification and diversification. A seed of mungbean is highly nutritious containing 24–28% protein, 1.0–1.5% fat, 3.5–4.5% fibre, 4.5–5.5% ash and 59–65% carbohydrates on dry weight basis and provide 334–344 kcal energy. Mungbean protein is considered to be easily digestible. Mungbean are tropical grain legumes widely grown in the sub-tropical countries of South and Southeast Asia. Nevertheless, these crops are cultivated over a wide range of latitudes in the regions where average diurnal temperatures during the growing season are warmer than about 20°C.

Keywords: mungbean, genetics, plant breeding, constrain, biotechnological tools

1. Introduction

Mungbean (*Vigna radiata* L. Wilczek) is cost-effectively most important crops of the pulse group. The mungbean is also known as green gram, golden gram and moong. Mungbean belonging to the subgenus *Ceratotropis* is a diploid species with 2n = 2x = 22 chromosomes. Worldwide, this crop is of minor importance with restricted geographical distribution, and has cautiously been subjected to detailed and intensive genetic and cytogenetic investigations. These legumes have a strategic position in Southeast Asian countries for nutritional security and sustainable crop production. Due to rich in quality protein, minerals and vitamins, they are inseparable ingredients in the diets of a vast majority of Indian population. When supplemented with cereals, they provide a perfect mix of essential amino acids with high biological value. Mungbean has the ability to fix atmospheric nitrogen in symbiotic association with Rhizobium bacteria, which enables them to meet their own nitrogen requirement and also benefit the succeeding crops [1]. These crops have also been reported to smother weed flora appreciably (20–45%) when intercropped with tall cereals and consequently, reduce the cost incurred on weed control [2]. On account of short duration and photo-thermo insensitivity, they are considered excellent crops for crop intensification and diversification. A seed of mungbean is highly nutritious containing 24–28% protein, 1.0–1.5% fat, 3.5–4.5% fibre, 4.5–5.5% ash and 59–65% carbohydrates on dry weight basis [3] and provide 334–344 kcal energy [4]. Mungbean protein is considered to be easily digestible. The dried grains of mungbean can be split or eaten whole after cooking and made into a soup or dhal. The iron availability in mungbean improves substantially to 7.2–11.3% through cooking practices such as soaking, fermenting and sprouting [5]. Mungbean is also widely relished as sprouts. The germinated grains have higher nutritional value as compared with asparagus or mushroom [6, 7]. Green pods and seeds can be cooked as vegetables. These pulses are frequently fed to children, convalescents and geriatrics or used when "breaking" a long fasting period owing to their ease of digestibility. The haulms are used for fodder and the beans husks and small broken pieces are useful as a feed concentrate. The crops are also grown for hay, green manure and cover crop. Mungbean makes better hay than urad bean as the stems and leaves are less hairy.

2. Origin and domestication

Mungbean is of Indian origin as is evidenced by their occurrence at archeological sites in the continent. *Vigna radiata* is native to north eastern India-Myanmar regions of Asia [8, 9]. *Vigna radiata var. sublobata* Verdc. is the closest wild relatives of the cultivated mungbean, respectively, and are regarded as their putative progenitors [10–13]. Based on morphological evidence alone, many researchers have considered var. sublobata as progenitor of mungbean [14, 15]. The detailed morpho-chemotaxonomic studies on wide collections of var. sublobata [12, 13, 16, 17], cross ability and chromosome pairing evidence [11, 18–20] have led to the conclusion that var. sublobata is a polymorphic taxon; two distinct morphological groups of it are the wild progenitors of mungbean and can be designated as *V. radiata* var. sublobata. The wild collections have characteristics conducive to domestication like annual growth habit, erect plant type, photoperiod insensitivity, more pods with high number of seeds, and smaller leaves. The present day cultivars of mungbean might have originated from new combinations of the already existing variants, changes in growth habit and seed size have been brought about by possible accumulation of recessive mutant genes [21]. Moreover, during domestication the dehiscent nature of pods and seed hardness of the wild progenitors have been selected out.

3. Ecology

Mungbean is tropical grain legumes widely grown in the sub-tropical countries of South and Southeast Asia [22, 23]. These legumes are grown at low to intermediate elevations on rainfed ecology. They perform best on good loamy soils with a well distributed rainfall of 750–900 mm per year, but are reasonably resistant to drought and susceptible to water logging. Mungbean is grown in kharif, winter and spring/summer seasons in different agro-ecological regions. Mungbean is grown as sole crops or as intercrops with sugarcane, maize, pearl millet, cotton, groundnut, sorghum and pigeonpea during kharif, as sole relay crop in rice fallows during winter and a sole catch crop during spring/summer seasons.

4. Historical perspective

Pure lines continued even during the early 40s to mid from these landraces were isolated on the basis of colour of stem, foliage, flower, unripe and ripe pods, seed colour and texture and other morphological features. The types selected were best suited in their respective regions mostly under low management. In mungbean, the first promising variety released was Type 1 for cultivation in Uttar Pradesh in 1948. It is a local selection from Muzaffarpur (Bihar). A large number of varieties were developed afterwards through selection from local materials and were released in different states between 1948 and 1970. Some of the important varieties developed through selection are Co 1 and ADT 1 (Tamil Nadu), Jalgaon (Maharashtra), Khargone 1, Krishna (Madhya Pradesh), and G 65 (Punjab). Jawahar 45 (Hybrid 45) released in 1971 in Madhya Pradesh and Type 44 released in Uttar Pradesh in 1972 was perhaps the first varieties developed through hybridization. Afterwards, a number of varieties were developed mainly through selection from the germplasm [24, 25]. Earlier a variety Virat has been released from ICAR—Indian Institute of Pulses Research, Kanpur, Uttar Pradesh; it is mature by 55 days only and fitted to rice wheat cropping system. This has resulted in development of appropriate production technologies and improved varieties besides basic knowledge on these crops.

5. Genetics

Information on the genetics of unusual traits in a crop is crucial for its systematic breeding programmes. Several studies have been conducted to know the genetics of qualitative and quantitative traits in addition resistance to major diseases and insect pests in mungbean, the results of which are presented in **Table 1**.

Qualitative traits	Characteristic	Gene involved	References
Plant type and growth habit	Erect, semi erect, semi-spreading or twining type	Single dominant gene T	[26, 27]
	Twining habit, semi-spreading habit is dominant to erect habit	Single dominant gene	[28]
	Erect, semi erect, semi-spreading or twining type	Dwarf mutant	[29]
	Indeterminate growth habit which inherited independently from leaf shape	Single dominant gene	[30]

Qualitative traits	Characteristic	Gene involved	References
Pigmentation	Purple hypocotyl which is dominant over green hypocotyl	Single gene 'A'	[31]
	'P' gene for the purple hypocotyl and a multiple allelic series 'C', 'C' and 'c' for purple, purple spotted and green epicotyl	Single dominant gene	[32]
	Anthocyanin pigmentation in hypocotyl, epicotyl, stem, petiole and peduncle	Single dominant genes	[28, 33, 34]
	A gene 'R' that conditions red colour of the cotyledons, hypocotyls and top of the leaflet stalk	Single dominant genes	[33]
	Anthocyanin pigmentation in the hypocotyl, epicotyl, stem, petiole, and peduncle	Single recessive gene	[35]
	Anthocyanin in hypocotyl	Two supplementary genes, designated as 'Sh' and 'Ph' with recessive epistatic interaction	[36]
	Purple pigmentation on stem, petiole and veins of leaves	Single dominant gene 'Pppl' with pleiotropic effect	[37]
Stem fasciation	Stem fasciations on the number of floral organs	Single recessive gene 'fsl' with a pleiotropic effect	[38]
Leaf traits	Inheritance of leaf size revealed that large leaflet is dominant over small leaflet	Single dominant gene	[37]
	Pentafoliate leaf	One recessive gene	[39]
	Pentafoliate leaf	Two recessive genes with duplicate gene action	[40]
	Nine foliate leaf mutant	Single recessive gene	[41]
	Induced unifoliata and multifoliata leaf mutants	Single recessive genes	[42]
	Narrow lanceolate leaf	Two recessive genes, 'nil' and 'n12'	[37]
	Lobed trifoliate leaf is dominant over entire leaf	Single dominant gene	[39, 43]
	Trilobite leaf	Two dominant genes 'Tibl' and 'T1b2' with duplicate action	[44]
Chlorophyll mutants	Chlorophyll mutants have been reported in mungbean with lethal and nonlethal effects. The albino seedling is controlled by monogenic recessive inheritance for the induced xantha, variegata, and greenish yellow chlorina mutants	Single recessive 'al' and 'I' genes	[42]
	Independent monogenic recessive inheritance for albina, chlorina, xantha and virescens types of chlorophyll mutants	Single recessive gene	[45, 46]

Qualitative traits	Characteristic	Gene involved	References
Inflorescence type	Simple inflorescence is governed by two dominant genes ('I1', '12') and double recessive homozygous genotype results in the compound inflorescence	Two dominant genes ('I1', '12')	[32]
	Inheritance of the number of clusters per node shows that a single dominant gene 'C' conditions one cluster per node and its recessive counterpart 'c' determines three clusters per node	Single dominant gene	[47]
	Induced sterility	Single dominant gene	[48, 49]
	A flower mutant with extended stigma and male sterility	Monogenic recessive inheritance	[50]
Flower colour	Four colours of the standard petal namely, red yellow, olive yellow, yellowish olive and light yellowish olive	Single dominant gene	[31]
	Light yellowish olive colour is partially dominant to olive yellow	Single partially dominant gene with gene symbols of 'Pg', 'Pb'	[51]
Pubescence	Dense plant pubescence	Single dominant gene 'Dp'	[51]
	Brown colour of the trait is recessive to colourless and therefore, dominant forms of both the genes are required for colourless pubescence	Two genes 'N' and 'Br'	[26]
	Pod pubescence is dominant over non-pubescence	Single dominant gene	[52]
Pod colour	A gene responsible for flower colour also conditions the colour of unripe pods	Single dominant gene	[31]
	Purple colour on the suture of unripe pod	Single dominant gene	[28]
	Inheritance of dry pod colour for light popcorn and almond biscuit colours	Genes 'lp' and 'lab'	[26]
	Colour of mature pods	Single dominant gene with black dominant over light brown colour	[28]
	Swollen tip is dominant over tapering pod tip	Single dominant gene 'Tp'	[26]
Pod shattering	Pod shattering is dominant to non-shattering	A single gene	[53]
	Resistance to shattering in the interspecific hybrids between mungbean and urd bean was dominant but nonshattering plants could not be recovered in the segregating generations suggesting that the pod shattering is a quantitatively inherited trait	A single gene	[54]

Qualitative traits	Characteristic	Gene involved	References
Seed coat colour	Thickness of the texture layer in seeds is under the quantitative genetic control while inheritance of the brown pigment in the texture layer The presence of brown pigment being dominant to its absence	Two complementary genes	[55]
	Inheritance of mottling in the seed coat is monogenic. The presence of anthocyanin being dominant to its absence. It indicated that the inheritance of black and green seed colours was controlled by a single gene, 'B' with black being dominant over green	A single gene	[56]
	Seed coat colour	A single gene	[57]
	Seed coat colour	Two independent dominant genes	[31]
	Seed coat colour The dominant alleles, 'A' and 'Sp', condition green and spotted seed coat whereas their recessive counterparts confer yellow and non-spotted colours	Two gene pairs	[33]
	Seed colour	Two genes, 'Dgsml' and 'Dgsm2'	[37]
	Each gene conditioning blue sap colour, buff sap colour and green chloroplast, respectively which together define the seed coat colour	Three gene pairs, 'Br', br and 'G'	[26]
	Seed coat colour	Three-gene model	[58]
	Seed coat colour	Three genes with several modifiers giving mottling patterns on yellow (mmbbgg), yellow green (mmBBgg), green (mmBBGG) and black (MMBBGG) seed coats	[59]
	Seed coat colour Black, brown, green mosaic, yellow mosaic, amber, green, and yellow seed coat	Four-gene (W, M, 'Br and G)	[60]
	Inheritance of seed coat colour	Five major genes with non-allelic gene interactions	[39]
Seed coat surface	Dull rough seed surface is monogenically dominant over glossy smooth surface and the gene symbols assigned for dull seed coat are 'C'	Single dominant gene	[26, 31, 33, 61]
	Digenic duplicate interaction (D1 and D2) is involved in the inheritance of seed luster, dullness being dominant over shiny	Two dominant gene	[41]

Qualitative traits	Characteristic	Gene involved	References
Cotyledon colour	Green cotyledon is conditioned by which is inherited independently of the red colour present in the hypocotyl and petiole	Single recessive gene 'gc'	[61]
Hard seededness	Hard seededness	Single dominant gene, 'Hdl'	[62]
	Hard seededness	Four QTL	[63]
Photoperiod response	The photoperiod insensitiveness is reportedly dominant over photo-sensitiveness	A single gene	[64]

Table 1.
Genetics of qualitative traits.

6. Production constraints

Traditionally, mungbean has been grown during kharif season. Development of short duration and disease resistant varieties has led their cultivation during spring/summer season in North and central India and during winter (rice fallows) in the coastal peninsula. The major constraints in achieving higher yield are lack of exploitable genetic variability, absence of suitable ideotypes for different cropping systems, poor harvest index, and susceptibility to biotic and abiotic stresses, besides non-availability of quality seeds of improved varieties. The major yield limiting harriers are lack of seedling vigour, excessive flower production, flower drops, poor pod setting, poor harvest index, monocarpic senescence, low response to inputs, narrow adaptation, indeterminate growth habit, staggered maturity and sensitivity to photoperiods and temperature. The phenomenon of compensation among yield components is considered to be main yield limiting factor. Limited variability has been exploited in varietal development programmes of these crops. Pedigree analysis of the released cultivars indicated that a small number of parents with high degree of relatedness were repeatedly used in crossing programmes. Diseases and insect pests cause considerable yield losses to mungbean. Mungbean yellow mosaic virus (MYMV), cercospora leaf spot (*C. canescens*, *C. cruenta*) and powdery mildew (*Ertisiphe polygoni* DC) are of considerable economic importance. Mungbean yellow mosaic virus and leaf crinkle during kharif and mungbean yellow mosaic virus during spring in North India and powdery mildew during winter season in coastal peninsula are the major diseases. During the vegetative stage, defoliators like hairy caterpillars, semi-looper and caterpillar are the common pests. Activity of thrips starts at the bud stage and poses serious problems when the crop attains peak flowering, resulting in heavy flower drop. There is no resistant variety against these insect pests. Pre-harvest sprouting especially in mungbean poses a serious threat to timely sown crop during rainy season. Intense heat and hot winds during May-June lead to flower drop and poor pod set in spring/summer crop.

7. Research needs

Varieties developed in the past with resistance to single stress may not be a viable solution as new diseases and insect pests are emerging. Therefore, varieties having resistance to more than one stress provide greater insurance. For mungbean, high yielding cultivars with crop duration of 85–90 days for kharif season and

65–70 days for spring season combining determinate growth habit, high harvest index and reduced photoperiod sensitivity are required. For summer cultivation, extra early varieties of 55–60 days with synchronous maturity are desirable. Vegetative growth should terminate with flowering and assimilates should be transported to developing pods [65, 66]. Recently, large seeded varieties like Pusa Vishal, SML 668, TMV 37, etc., with early and synchronous maturity have been developed which have great market demand. To break the yield plateau in mungbean, there is a need to develop suitable plant type for target environments. In high input cereal-cereal systems, mungbean has to fit in gaps. For this, plant type that is determinate, photo-thermo insensitive, early maturing and high yielding with high harvest index needs to be developed. Good seedling vigour, distinctive vegetative and reproductive phases and high harvest index will be essential components of this plant type. There is good scope to utilize wild and cultivated *Vigna* species to incorporate novel characters and broaden the genetic base.

8. Molecular diversity analysis

Assessment of genetic diversity using RAPD analysis shows close similarity among mungbean cultivars [67]. The study reveals narrow genetic base of Indian cultivars probably due the repeated use of limited ancestors in their pedigrees. This observation has further been confirmed using RAPD [68, 69] and ISSR [70, 71] markers. Amplified fragment length polymorphism (AFLP) markers have also been used in mungbean to test their usefulness in genetic diversity assessment [67]. The long primers yielded significantly higher number of discrete and detectable bands as well as polymorphic bands than 10-base primers. The results show that long primers can be used efficiently for analyzing genetic diversity and the relationships in mungbean germplasm.

9. Mapping of genes/QTLs

Increasing the seed weight has been one of the major objectives to develop high yielding varieties. Molecular markers are now available which are linked to orthologous seed weight loci. RFLP markers to locate major QTLs for orthologous seed weight in mungbean. They found that the genomic regions in cowpea and

Characteristic	Marker	Genes/QTLs/remarks	References
Seed weight	RFLP	Major QTLs	[72]
	RFLP	Suggested a weak association between seed weight and hard seededness in mungbean by analyzing a F_2 population	[62]
	RFLP	Four loci for hard seededness and 11 loci	[63]
Powdery mildew	RFLP	Genes, '13 m' and 'Thiz2' identified in a cross VC3890 × TC1966	[65]
	RFLP	Two QTLs, '13MR1' and 'PMR2' have been identified	[73]
	RFLP	A single locus has been identified that explains 86% of variation associated with resistance to powdery mildew in mungbean	[74]

Table 2.
Genes/QTLs/remarks of important traits.

mungbean that have the major effect on seed weight span the same RFLP markers in both the species. These markers are co-linear in arrangement on homologous linkage groups in both the crops. Attempts to breed large and hard seeded varieties of mungbean have not been very successful because of negative genetic correlation between these two traits as a result of pleiotropy or genetic linkage. Studies on the genetic relationship between hard seededness and seed weight, however, are not conclusive. QTL mapping approach using molecular markers have been employed to investigate the linkage relationship between these two traits (**Table 2**).

10. Breeding approach

In order to develop high yielding disease resistant varieties in mungbean, the common breeding methods employed were pure line selection, hybridization followed by pedigree selection, mutation breeding and wide hybridization. While exercising selection, major emphasis has been placed on short duration, photo and thermo-insensitivity, synchronous maturity and resistance to mungbean yellow mosaic virus and powdery mildew. More than 150 varieties have been developed in India employing pure line selection, pedigree method of selection following hybridization, mutation and wide hybridization. The first variety of mungbean was Type 1 developed from local selection of Muzaffarpur (Bihar), which has been extensively been used afterward as one of the parents in hybridization programmes for the development of improved varieties like Type 2, K 851, T 44 and Sunaina. Utilization of T 44 in hybridization has resulted in the development of Pusa Baisakhi which, in turn, has given PIMS 4 and Jyoti. Through mutation breeding, about 14 varieties using gamma rays or occasionally ethyl methane sulphonate as mutagens have been developed. Varieties developed through mutation like CO 4, Pant Moong 2, TAP 7, BM 4, MUM 2 and TARM 1.

10.1 Parental selection

The main reason that the expected yield advances by the conventional component breeding methods have not materialized in mungbean is that the parents used in crossing programmes are not duly evaluated before their use. Seed yield of parents has a positive significant bearing on the yield of the progenies and in the inheritance of this character, additive variance is of paramount importance than the nonadditive variance, although many a times the latter also has significant bearing. The choice of the parents besides on their agronomic attributes like yield and its components must also be based on their genetic diversity, phenotypic stability and combining ability. So logically all the would-be parents must be evaluated by their progeny tests across environments and locations before their use in a crossing programme. A progeny test provides genetic composition of the parental plants and helps in selection of superior ones. In self-pollinated crops like mungbean, many minor genes of additive effect control yield and in any breeding programme, the ultimate goal is to accumulate and harness these genes. High yielding varieties from different genetic backgrounds and carrying different genes for yield when crossed and subjected to replicated progeny tests are expected to generate higher frequency of high yielding plants. Yield stability in mungbean is very important owing to significantly variable response of high yielding varieties across locations and years. Work on stability analysis done in mungbean shows that no high yielding varieties are stable across time and space. All the potential parents in a hybridization programme must be evaluated for their mean yield performance and yield stability, F1 performance, F2 mean yield and the variance generated, combining ability and their interaction with the important environmental variables. All these variables give a measure of the comparative potential of different F2 crosses. It is desirable that the progenies of only those parents be advanced beyond F2 generation that show high grain yield, yield stability, a positive general combining ability for grain yield and that are of distant genetic origin. Progenies of parents with low yield and negative general combining ability for yield must not be advanced beyond F2 generation. In an intra-species crossing programme, one parent should be a good agronomic base with higher stability and the other parent a good general combiner for yield and its components. Crosses with this strategic selection of parents are expected to give a wide range of genetic variability. To achieve stability and get a true measure of inherent genetic potential, the parental lines must be tested over a number of locations and get their combining ability estimates.

10.2 Component breeding

Fifty years of conventional approach of engineering different yield components in mungbean to build up a new plant type with higher productivity levels has thus far given only modest yield gains over the traditional cultivars. This approach has failed to break the present yield barriers as a whole and bring changes of scale. Based upon correlation analysis of various yield components, selections have been based mainly on number of pods per plant, seeds per pod and 100-seed weight and sometimes also on number of pods per bunch and branches per plant or podding per unit area usually called 'Pod Index'. Pods per plant is by far the most important yield component and almost all the workers have found it having positive correlation with seed yield. It is the best selection index for seed yield and could be increased by increasing number of branches per plant or number of bunches per plant or by increasing the number of pods per bunch. Most of the work has shown that branch number per plant is negatively correlated with seed yield, but bunches per plant has mostly been found to be positively correlated with seed yield. Ramanujam [75] and others have found that pods per bunch and bunches per plant both are positively correlated with yield. Increasing pods per bunch is physiologically constrained in grain legumes owing to fall of flowers and unripe or partially filled pods. It seems the most feasible path to increase seed yield is through increasing number of bunches per plant. This in essence means a plant with more number of nodes with a shorter internode length, with three to four erect branches emerging from the lower to lower-middle nodes at around 20–30° angle with the main stem, and sympodial bearing of pod inflorescences coming from the upper nodes of the main stern, each carrying around 8–10 pods. Number of seeds per pod has been shown mostly positively correlated with seed yield but many workers show it to be negatively correlated with yield. However, an optimal level of 14–16 seeds per pod should be a breeders objective. Seed weight is generally negatively correlated with seed yield but some results have shown it to be positively correlated. The strength of the newly developed second generation varieties like Pusa Vishal and Pusa Ratna lies in the fact that they have more seeds per pod (12–16) with higher 100-seed weight (5.0–5.5 g) without compromising on the pod number per plant. Many researchers advocated cereal mimics with sympodial bearing and suggested increasing pods per plant through the path of increasing the average number of pods per node and building up a soybean like plant type in mungbean [76]. He found the main stem bearing under the control of a single recessive gene and normal conventional bearing to the incompletely dominant. Plant height has been found positively correlated with seed yield. An optimal upright plant height incorporates more functional nodes and thus more number of pods per plant. After pods per plant, this is the second most important character to be used for selection of seed yield. Owing to their high heritability, 100-seed weight and branch number could be excellent selection criteria but for their unfavorable correlations with yield. Also due to the compensatory mechanisms operating within the plant as a whole, this correlation based selection methodology has not brought the desired productivity levels in mungbean. Alternatively, the best option is direct selection for seed yield on a unit area basis.

10.3 Conventional breeding methods

Most of the high yielding varieties of mungbean bred and released so far have been developed through single cross pedigree method of selection. The single plant selections made in the early generations restrict carrying forward the bulk of created variability, which gets lost quickly giving way to homozygosity with each succeeding generation. This method has served the mungbean improvement programme well in the past, but lately no productivity advances are materializing due to the inherent genetic limitation of the method. The intermitting of selected F_2 plants and selections in the late generations will help to harness most of the desirable genes.

10.4 Early generation testing

The early generation yield trials allow early identification of better performing crosses and F, derived lines within the individual crosses. However, selection emphasis is given in later generations only. The F_2 derived family selection is very appropriate in mungbean, which is prone to high GE interactions and low seed increase ratio, which renders pedigree, bulk and single seed decent methods inefficient. The time required is less and emphasis is on grain yield in replicated progeny tests. It was developed in Canada as a modification of bulk method. The F2 derived family selection takes benefits of early generation yield testing to eliminate efficiently all the undesired materials both between and within the hybrid populations. Replicated yield trials are conducted across locations/environments for early generation selections among and within populations. These selections are further evaluated and final selections for high yield are made in only the best of families or populations. Due emphasis must be given to make site specific selections for different agro-climatic and production systems. Also depending upon the demands of the location and system, input responsiveness of the selections under high management conditions may be tested.

10.5 Mutation breeding

Both physical and chemical mutagens have been employed in improvement of mungbean crop India. The main drawbacks of this method are that the frequencies of desirable mutants are very less, necessitating evaluation of very large population and the difficulty in identification and scoring of micromutations. Tickoo and Chandra [77] using both physical and chemical mutagens could induce significantly higher variability in mungbean for characters like yield per plant, pods per plant, seed number per pod, seed weight, days to flower and harvest index in M_2, generation. Mean values of all the characters had a negative shift in M_2 but after selection changed to positive direction in M_3 but were still associated with significantly higher interfamily and overall variances than the control populations. Such characters may be incorporated into the cultivated varieties by backcross method. Many a times the selected mutants have been released as new varieties as such for cultivation. Some varieties have been released in India including Pant Mung 2, Co 4, Dhauli, TAP 7, BM 4 and MUM 2 and some in Pakistan including NM 51 and NM 54. The latter two large seeded varieties resistant to MYMV have been developed by hybridization and irradiation of the F_1 seeds.

11. Thrust areas

Resistance to biotic stresses *namely*, mungbean yellow mosaic virus (MYMV), powdery mildew (PM), cercospora leaf spot (CLS), root diseases caused by Pythium spp. There is variability in the virulence of MYMV in white fly for incubation of different isolate in mungbean plant for genes governing resistance/tolerance to the virus and tim vector. Gene transfer across species in Vigna group has been found difficult. Biotechnological tools are being presently used to overcome these barriers. In the field, spreader row technique has been found to be most effective in screening the materials. However, parental lines to be used in crossing and promising selections before their release for cultivation may she screened under artificial inoculation.

12. Conclusions

Mungbean has the distinct advantage of short crop duration. This fact together with its atmospheric nitrogen fixing ability makes it an indispensable component in various cereal-based cropping systems. Mungbean has tremendous scope for horizontal expansion subject to some committed research inputs to overcome its various productivity bottlenecks. Proper evaluation and utilization of germplasm from secondary and tertiary gene pools by conventional and biotechnological tools immediate priorities. Incorporation of genes from its closely related species tor resistance or tolerance to biotic factors like MYMV, bruchids and abiotic factors like sensitivity to photoperiods, high temperatures, drought, waterlogging, pre-harvest sprouting and nutrient use efficiency' and response to irrigation must be accomplished. Breeding for efficient fixing of atmospheric nitrogen has to be priority for higher response to the applied nitrogen in recombination breeding. Mungbean is grown in different seasons and different cropping systems necessitating the development of varieties of different maturity span. The varieties for wheat based cropping systems of Indo-Gangetic-Plains should have crop duration of 60 days with average yield of 1.3–2.0 tonnes per ha.

Acknowledgements

Authors are highly thankful to the Vice chancellor M.G.C.G.V. Satna, Chitrakoot, Madhya Pradesh for providing all the necessary facilities to write this book chapter and also acknowledge to ICAR-IIPR, Kanpur for their support.

Conflict of interest

There is no conflict among the authors.

DOI: http://dx.doi.org/10.5772/intechopen.85657

Author details

Suhel Mehandi[1*], Syed Mohd. Quatadah[1], Sudhakar Prasad Mishra[2], Indra Prakash Singh[1], Nagmi Praveen[3] and Namrata Dwivedi[2]

1 ICAR—Indian Institute of Pulses Research, Kanpur, Uttar Pradesh, India

2 Mahatma Gandhi Chitrakoot Gramodaya Vishwavidhyalaya, Satna, Madhya Pradesh, India

3 Rajmata Vijayaraje Scindia Krishi Vishwavidyalaya, Gwalior, Madhya Pradesh, India

*Address all correspondence to: suhelgpb@gmail.com

References

[1] Ali M. Effect of summer legumes on productivity and nitrogen economy of succeeding rice (*Oriza sativa*) in sequential cropping. Indian Journal of Agricultural Sciences. 1992;**62**:466-467

[2] Ali M. Weed suppressing ability and productivity of short duration legumes with pigeonpea under rainfed condition. Tropical Pest Management. 1988;**34**:384-387

[3] Tsou CS, I Isu MS, Tan ST, Park HG. Protein quality of mungbean and its improvement. Aito Horticulturac. 1979;**93**:279-289

[4] Srivastava RP, Ali M. Nutritional Quality of Common Pulses. Vol. 65. Kanpur: Indian Institute of Pulses Research; 2004

[5] Yang KY, SCS T, Lee TC. Effect of cooking on in vitro iron bioavailability of various vegetables. In: Lee TC, Ho CT, editors. Bioactive Compounds in Foods: Effect of Processing and Storage. Washington DC, USA: American Chemical Society; 2002. pp. 130-142

[6] Tan SY. In: Summerfield RJ, Roberts EH, editors. Grain Legume Crops. London, UK: Collins; 1973. pp. 20-26

[7] Mehandi S, Singh IP, Bohra A, Singh CM. Multivariate analysis in green gram [*Vigna radiata* (L.) Wilczek]. Legume Research. 2015;**38**(6):758-762

[8] de Candolle A. Origin of Cultivated Plants. New York, USA: Hafner; 1988

[9] Ghafoor A, Sharif A, Tahir M. Evaluation of black gram germplasm. Pakistan Journal of Botany. 1886;**30**:227-238

[10] Singh BV, Ahuja MR. *Phaseolus sublobatus* Rox: A source of resistance to yellow mosaic virus for cultivated mung. Indian Journal of Genetics & Plant Breeding. 1977;**37**:130-132

[11] Lukoki L, Marechal R, Otoul E. The wild ancestors of the cultivated beans *Vigna radiata* (L.) Wilczek and *V. mungo* (L.) Hepper. Bulletin of Jardin Botany National Belgium. 1980;**50**:385-391

[12] Miyazaki S, Kawakami J. Differences in several characters among mungbean, black gram and their wild relatives. Japanese Journal of Tropical Agriculture. 1981;**25**:1-5

[13] Chandel KPS. Role of wild *Vigna* species in the evolution and improvement of mung (*Vigna radiata* L. Wilczek) and urdbean (*V. mungo* L. Hepper). Annals of Agricultural Research. 1984;**5**:98-111

[14] Roxburghi W. Flora Indica. Vol. 763. New Delhi, India: Today and Tomorrow's Printer and Publishers; 1832

[15] Cooke T. The flora of the Presidency of Bombay. In: Botanical Survey of India. Vol. I. 1903. pp. 399-404

[16] Miyazaki K. Classification and phylogenic relationship of the *Vigna radiata* mungo sublobata complex. Bulletin of National Institute of Agricultural Sciences, Japan. 1982;**33**:1-61

[17] Arora RK. Diversity and collection of wild *Vigna* species in India. FAO/IBPGR Plant Genetic Resources Newsletter. 1985;**63**:26-33

[18] Biswas MR, Dana S. Black gram x rice bean cross. Cytologia. 1975;**40**:787-795

[19] Karmarkar PG. Cytogenetics of *Vigna* species hybrids and allopolyploids [Ph.D. thesis]. Kalyani, India: Bidhan Chandra Krishi Vishwavidyalaya; 1985

[20] Karmarkar PG, Dana S. Cytogenetic identification of *Vigna* sublobata collection. Nucleus. 1987;**30**:47-53

[21] Sen NK, Murty ASN. Inheritance of seed weight in green gram (*Phaseolus aureus* Roxb.). Genetics. 1960;**45**:1559-1562

[22] Rachie KO, Roberts LM. Grain legumes of lowland tropics. Advances in Agronomy. 1974;**26**:1-132

[23] Konarev AV, Tomooka N, Vaughan DA. Proteinase inhibitor polymorphism in the genus *Vigna* subgenus Ceratotropis and its biosystematic implications. Euphytica. 2002;**123**:165-177

[24] Singh DP, Satyanarayana S. Creating higher genetic yield potential in urdbean. In: Recent Advances in Pulses Research. Kanpur, India: Indian Society of Pulses Research and Development; 1997. pp. 157-171

[25] Mehandi S, Mishra SP, Tripathi RC, Singh IP. Genetic variability, heritability and genetic advance for yield and its related traits in mungbean [*Vigna radiata* (L.) Wilczek] genotypes. International Journal of Current Microbiology and Applied Sciences. 2018;(Special Issue-7):3818-3824

[26] Siddique M, Malik MFA, Shahid IA. Genetic divergence, association and performance evaluation of different genotypes of mungbean. International Journal of Agriculture and Biology. 2006;**8**(6):793-795

[27] Khattak GSS, Ashraf M, Haq MA, Srinives P. Genetic architecture of seed yield and yield components in mungbean [*Vigna radiata* (L.) Wilczek]. Tropical Agriculture. 2002;**79**:260-264

[28] Pathak H, Singh S. Inheritance studies in green gram. Indian Journal of Genetics & Plant Breeding. 1963;**23**:215-218

[29] Pande K, Raghuvanshi SS. Gamma rays induced high yielding dwarf mutant in *V. radiata* L. Wilczek. Mutation Breeding Newsletter. 1988;**32**:6-7

[30] Talukdar T, Talukdar D. Inheritance of growth habbit and leaf shape in mungbean [*Vigna radiata* (L.) Wilczek]. Indian Journal of Genetica & Plant Breeding. 2003;**63**:165-166

[31] Bose RD. Studies in Indian pulses. IX. Contributions to the genetics of mung (*Phaseolus radiaus* L.). Indian Journal of Agricultural Sciences. 1939;**9**:575-594

[32] Sen NK, Ghosh AK. Genetic studies in green gram. Indian Journal of Genetics and Plant Breeding. 1959;**19**:210-227

[33] Rheenen V. The inheritance of some characters in the mungbean, *Phaseolus aureus* Roxb. Genetica. 1965;**36**:412-419

[34] Mishra RC, Sahu RC, Tripathi D. Inheritance of hypocotyl color in green gram (*Phaseolus aureus* Roxb). Indian Science Congress Association Proceedings. 1970;**57**:290

[35] Apparao S, Jana MK. Inheritance of anthocyanin coloration in *Phaseolus* mutants. Indian Science Congress Association Proceedings. 1973;**60**:302

[36] Mukherjee A, Pradhan K. Genetics of mungbean: Anthocyanin pigment in hypocotyl. Journal of Interacademicia. 2002;**6**:434-437

[37] Dwivedi S, Singh DR. Inheritance of narrow trifoliate leaf in mungbean (*Vigna radiata* (L.) Wilczek). SABRA Journal. 1985;**17**:177-180

[38] Dwivedi S, Singh DP. Inheritance of fasciation in mungbean (*Vigna radiata* L. Wilczek). Indian Journal of Genetics & Plant Breeding. 1990;**50**:81-82

[39] Chhabra AK. Inheritance of lobed and pentafoliate leaf in mungbean. Indian Journal of Pulses Research. 1990;**3**:69-72

[40] Muralidharan V, Rangasamy SR, Chandrababu R, Rathnaswamv R. Pentafoliate leaf and its inheritance in mungbean. Indian Journal of Pulses Research. 1991;**4**:195

[41] Bhadra SK. Inheritance of nine-foliate leaf in mungbean (*Vigna radiata* (L.) Wilczek). SABRAO Journal. 1991;**23**:71-73

[42] Santos IS. Induction of mutations in mungbean (*Phaseolus aureus* Roxb.) and genetic studies of some of the mutants. 1969;**8**:169-179

[43] Pokle YS. Spontaneous mutation in mung (*Phaseolus aureus* Roxb.). Science and Culture. 1972;**38**:142

[44] Sareen PK. Further genetic analysis of the trilobate leaf mutants in mungbean (*Vigna radiata* var. aureus (L.) Wilczek). Current Science. 1985;**54**:930-931

[45] Bahl JR, Gupta PK. Chlorophyll mutations in mungbean (*Vigna radiata* (L) Wilczek). Theoretical and Applied Genetics. 1982;**63**:23-26

[46] Bahl JR, Gupta PK. Inheritance of two induced lethal chlorophyll mutations in mungbean. Current Science. 1984;**53**:147448

[47] Singh BB, Singh DP. Prospects of creating higher yield potential in chickpea. In: Asthana AN, Ali M, editors. Recent Advances in Pulses Research. Kanpur, India: Indian Society of Pulses Research and Development; 1995. pp. 65-88

[48] Ganguli PK. Radiation induced sterile mutant in green gram (*Phaseolus aureus* Roxb.). Indian Science Congress Association Proceedings. 1973;**60**:640-641

[49] Saini RG, Minocha JL, Singh A. Sterile mutants in *Phaseolus aureus*. Science and Culture. 1974;**40**:37-38

[50] Seenaiah P, Satyanarayana A, Rao YK, Naidu NV. Genetic diversity in the genus *Vigna* based on morphological traits and isozyme markers. SABRAO Journal. 1990;**35**:103-112

[51] Murty BK, Patel GJ. Inheritance of seed coat colour in green gram. Indian Journal of Genetics & Plant Breeding. 1972;**32**:373-378

[52] Khadilkar BT. Inheritance of two characters in green gram (*Phaseolus aureus* Roxb. Piper). Akola Agricultural College Magazine. 1963;**3**:29-32

[53] Verma, Krishi. Inheritance of some qualitative characters in green gream (*Phaseolus aureus* Roxb.). Indian Journal of Heredity. 1969;**1**:105-106

[54] Singh P, Singh IB, Singh U, Singh HG. Interspecific hybridization between mung (*Phaseolus aureus* Roxb.) and urd (*Phaseolu smungo* L.). Science and Culture. 1975;**41**:233-234

[55] Watt EE, Marechal R. The differences between mung and urd beans. Tropical Grain Legume Bulletin. 1977;**7**:31-33

[56] Chen HM, Liu XH. Inheritance of seed color and luster in mungbean (*Vigna radiata*). Hunan Agricultural Science & Technology Newsletter. 2001;**2**:8-12

[57] Lambrides CJ. Segregation distortion for seed testa color in mungbean (*Vigna radiata* L. Wilczek). The Journal of Heredity. 2004;**95**:532-535

[58] Singh. Inheritance of seed coat colour in *Phaseolus aureus* L. Indian Science congress Association Proceedings. 1973;**60**:321-322

[59] Laurel VB, Ramirez DA. Inheritance of seed coat color in mungbean, *Vigna radiata* (L.) Wilczek. Philippines Journal of Crop Science. 1994;**19**:3-19

[60] Mehandi S, Singh IP, Bohra A, Singh CM. (Supplement on Genetics & Plant Breeding) Estimates of genetic variability and heritability for yield and yield component traits in mungbean [*Vigna radiata* (L.) Wilczek]. The Bioscan. 2013;**8**(4):00-00

[61] Thakare et al. Green cotyledons mutant in mungbean [*Vigna radiata* (L.)Wilczek]. Current Science. 1980;**49**:945-946

[62] Lambrides CJ. Breeding for improved seed quality traits in mungbean (*Vigna radiata* L. Wilczek) using DNA markers [Ph.D. Thesis]. Australia Lambrides CJ: University of Queensland; 1996

[63] Humphry ME, Lambrides CJ, Chapman SC, Aitken EAB, Imrie BC, Lawn RL, et al. Relationships between hard seededness and seed weight in mungbean (*Vigna radiata*) assessed by QTL analysis. Plant Breeding. 2005;**124**:292-298

[64] Verma P. Inheritance of photosensetivity in mungbean (*Phaseolus aureus* Roxb.). Mysore Journal of Agricultural Sciences. 1971;**5**:477-480

[65] Sahli AD, Das K. Studies on growth and yield of three mungbean (*Vigna radiata* L. Wilczek) cultivars. Indian Journal of Plant Physiology. 1979;**22**:147-154

[66] Pawar SE, Bhatia CR. The basis for grain yield differences in mungbean cultivars and identification of yield limiting factors. Theoretical and Applied Genetics. 1980;**57**(1):71-1 75

[67] Ghat KV, Lakhanpaul S, Chadha S. Amplified fragment length polymorphism (AFLP) analysis of genetic diversity in Indian mungbean (*Vigna radiata* (L.) Wilczek) cultivars. Indian Journal of Biotechnology. 2005;**4**:56-64

[68] Betal S, Chowdhury PR, Kundu S, Raychaudhuri SS. Estimation of genetic variability of *Vigna radiata* cultivar by RAPD analysis. Biologia Plantarum. 2004;**48**:205-209

[69] Afzal MA, Hague MM, Shanmugasundaram S. Random amplified polymorphic DNA (RAPD) analysis of selected mungbean (*Vigna radiata* (L.) Wilczek) cultivars. Asian Journal of Plant Sciences. 2004;**3**:20-24

[70] Sinha RR, Das K, Sen SK. DNA markers in mungbean. India Embryoids from mesophyll protoplast of Vigna mungo (L) Hepper: A seed legume crop plant [Ph.D. Thesis]. Vol. 22. University of Mumbai; Basic Life Sciences. 2003. pp. 209-211

[71] Saini A, Reddy SK, Jawali N. Evaluation of long primers for AP-PCR analysis of mungbean (*Vigna radiata* (L.) Wilczek): Genetic relationships and fingerprinting of some genotypes. Indian Journal of Biotechnology. 2004;**3**:511-518

[72] Fatokun CA, Menancio-Hautea D, Danesh D, Young ND. Evidence for orthologous seed weight genes in cowpea and mungbean based on RFLP mapping. Genetics. 1992;**132**:841-846

[73] Chaitieng B, Kaga A, Han OK, Wang KW, Wongkaew S, Laosuwan P, et al. Mapping a new source of resistance to powdery mildew in mungbean. Plant Breeding. 2002;**121**:521-525

[74] McIntyre CL, Humphry ME, Magner T, Aitken EAB, Liu CJ. Identification of a major locus conferring resistance to powdery mildew (*Erysiphe polygoni* DC) in mungbean (*Vigna radiata*

L. Wilczek) by QTL analysis. Genome. 2003;**46**:738-744

[75] Ramanujam S. Genetic diversity, stability and plant type in pulse crops. In: Proceedings of the First International Mungbean Symposium. Shanhua, Taiwan: Asian Vegetable Research and Development Centre. 1975. pp. 167-176

[76] Satyanarayana A. Pulses in rice fallows. In: Proceedings of the International Workshop on Varietal Improvement of Chickpea, Pigeonpea and Other Upland Crops in Rice Based and Other Cropping Systems; March 19-22; Kathmandu. 1998. pp. 352-354

[77] Tickoo, Chandra. Mutagen induced polygenic variability in mungbean [*V. radiata* (L) Wilczek]. Indian Journal of Genetics & Plant Breeding. 1999;**59**(2):193-201

Chapter 6

The Productivity of Selected Species and Cultivars of Legumes Grown for Seeds in Organic Production System

Księżak Jerzy and Bojarszczuk Jolanta

Abstract

The aim of the study was to assess the yielding of selected legume species with diversified morphological structure cultivated for seeds in ecological system. The field experiment was carried out in 2016–2018. The first factor was legume species: faba bean, field pea, yellow lupine, and blue lupine, and the second factor was varieties of legumes: faba bean (Granit and Amulet), field pea (Hubal and Batuta), blue lupine (Kurant and Regent), and yellow lupine (Bursztyn and Perkoz). After the harvest, the grain yield of legume plants and the weight of a thousand seeds were determined. The plant structure was determined (length of the part of fruiting stem, number of pods and seeds per plant, number of seeds in the pod, number of fruiting nodes, number of pods and seeds from the node). In addition, the content of selected nutrients (protein, fiber, fat, macroelements) was determined in seeds. Studies showed that in ecological conditions, the pea cultivation, especially Hubal variety (with bipinnate leaves), enabled obtaining the largest seed yield, while the smallest seed yields yellow lupine independent of the morphological type. The self-completing varieties of faba bean, yellow lupine, and blue lupines were yielded at a higher level than varieties with a traditional growth type. Among the pea varieties assessed, the variety Hubal yielded better (with bipinnate leaves). Significantly, higher yield of protein is provided by faba bean cultivation, while the smaller level of pea and yellow lupine.

Keywords: cultivar, ecological system, legume, seeds, productivity

1. Introduction

Currently, there is a growing interest in growing leguminous plants, as their high fodder value, universal consumption values, and their role in a sustainable and ecological production system are more and more widely appreciated [1, 2]. An extremely important trait of these species is also the ability to bind atmospheric nitrogen (about 3–6 million tons per year by global crops), which allows to reduce CO_2 and NO emissions into the atmosphere and at the same time allows to reduce the demand for nonrenewable energy sources for food production [3]. Legume seeds and legume-based food are an important and sustainable source of nutrients for human diet, especially carbohydrates and proteins [4, 5]. They also contain active substances such as phenolic compounds whose antioxidant activity and health features are the subject of many studies [6, 7]. They are used to produce functional food and improve food nutritional value [8–10]. According to Duranti [10] and Vioque et al. [4], an increase in consumer awareness of the health benefits of these proteins can stimulate the production of legumes. In addition, their high protein and energy content make their seeds an excellent feed source [6, 11]. In addition, according to Doležal et al. [12], Fraszer et al. [13], and Szyszkowska et al. [14], protein, which has a significant influence on the results of animal production, is the nutrient that determines the nutritional value.

The role and importance of leguminous plants in agriculture, regardless of the production system, the increase in the area of organically cultivated agricultural land, and the increase in consumer knowledge concerning the health value of leguminous seeds, prompted us to undertake research evaluating the productivity of four legume species with diversified morphological structure of organically cultivated plants.

2. Materials and methods

The field experiment was carried out in the years 2013–2015, in a split-plot design, in four replications. The first factor was legume species: faba bean, field pea, yellow lupine, and blue lupine, and the second factor was varieties of legumes: faba bean—Granit (self-completing) and Amulet (traditional growth type), peas—Hubal (traditional growth type) and Batuta (with bipinnate leaves), blue lupine—Kurant (traditional growth type) and Regent (self-completing), and yellow lupine—Amber (traditional) and Perkoz (self-completing). Plant density was: faba bean (70 units·m^{-2}), peas, yellow lupine, and blue lupine (100 units·m^{-2}). The size of the plot, for harvest, is 22.0 m^2. The experiment was carried out on the soil of a very good rye complex, class IIIa. The content of available phosphorus (in mg per 100 g of soil) ranged from 10.15 to 11.8%, potassium from 11.1 to 20.7%, magnesium from 2.8 to 4.1%, and humus from 1.34 to 1.39%. Sowing was carried out from the 2nd to the 29th of April. The collection of pea and blue lupine was made at the complete maturity in the first days of August, and faba bean and yellow lupine in the second and third decade of August. For the purposes of care, the harrowing of legumes was performed twice. During the growing season, dates of the developmental phases of legumes have been recorded. Before the harvest, on ten random chosen plants from each plot, morphological features were determined (height of plants, height of the first pod, share of pods in the plant, number of pods per plant). After the harvest, the seed yield and the weight of thousand seeds per 14% were determined. The content of total nitrogen and phosphorus was determined in the seeds (control flow analysis (CFA), potassium content (emission atomic spectrometry), crude fat, crude fiber, and ash content (weight method). The significance of the influence of the experimental factors on observed characters was assessed using the analysis of variance, determining Tukey's half-intervals at the significance level of $\alpha = 0.05$.

3. Results and discussion

During the study period, there were significant differences in the growth and development of faba bean. In 2016, precipitation was fairly evenly distributed in individual months (**Table 1**). Despite the fact that the sum of precipitation during the growing season was much lower than the average for many years, it was conducive to yielding legumes. Low Sielianinow's coefficients (less than 1) were recorded in June 2017 and 2018 during the most intense demand of plants for water, i.e., during flowering and emergence of pods (**Table 1**). It can therefore be concluded that in the most critical period of faba bean development, plants in these years, and especially in 2017, were relatively poorly supplied with water. In addition, during this period higher air temperatures were noted, which also did not favor the field bean harvest. According to Podleśna et al. [15] and Faligowska [16], the amount and distribution of precipitation during the growing season of plants is one of the most important factors affecting the yield level of leguminous plants. According to Atkins and Smith [17], the shortage of precipitation combined with the high air temperature is particularly unfavorable during germination and setting of pods, because the plants shed flowers and pods, which in turn reduces the crop yield.

The weather conditions are shown by calculating the hydrothermal coefficient of water supply for individual years according to Sielianinov's index (K). The following formula was applied:

$$K{=}Moo \times 10$$
$$Dtt \times d$$

where K is the hydrothermal coefficient for individual months, Mo is the total monthly precipitation, and Dt is mean daily temperatures in a particular month.

The yield of legume seed significantly depended on the course of weather conditions during the growing season, the legume species, its type of growth and development (self-completing, traditional), or the type of foliage (with bipinnate leaves, traditional). The highest level of yields of all species was recorded in 2016, and they were higher by about 75% than in 2017 and by about 40% than in 2018 (**Table 2**). On average, for 3 years among the assessed species, the highest yields were provided

Specification	Month							Sum/mean
	III	IV	V	VI	VII	VIII	IX	
2016								
Rainfall (mm)	52.3	45.1	39.4	60.1	81.9	53.6	20.3	352.7
Temperature (°C)	3.9	9.2	14.9	18.7	19.2	18.1	15.7	14.2
Sielianinov's index (K)	4.23	1.75	0.85	1.07	1.37	0.95	0.43	1.52
2017								
Rainfall (mm)	35.8	69.1	34.4	32.6	86.3	55.3	102.7	416.2
Temperature (°C)	5.7	7.5	13.9	18.1	18.6	19.6	13.9	13.9
Sielianinov's index (K)	2.02	3.07	0.80	0.60	1.49	0.91	2.46	1.62
2018								
Rainfall (mm)	14.1	25.3	97.4	44.6	118.5	70.6	—	370.5
Temperature (°C)	—	13.3	17.0	18.4	20.4	20.2	—	17.9
Sielianinov's index (K)	—	0.63	1.85	0.81	1.86	1.13	—	1.26
Average rainfall from many years*	34	50	67	79	87	71	58	63.7
Average temperature from many years (°C)	2.1	8.0	13.6	16.8	18.5	17.8	13.2	14.3

**Mean for the years 1961–2017.*

Sielianinov's index: < 0.5, drought; 0.5–1.0, semi-drought; 1.0–1.5, border of optimal moisture; > 1.5, excessive moisture.

Table 1.
Course of weather conditions during the vegetation.

Legume species	Cultivar	Seed yield			Weight of 1000 seeds		
		2016	2017	2018	2016	2017	2018
Blue lupine	Kurant	2.53	1.41	2.15	142.5	142.3	147.2
	Regent	2.68	1.34	2.26	138.2	128.6	131.4
Yellow lupine	Bursztyn	1.55	0.85	1.02	104.8	130.7	133.2
	Perkoz	1.66	1.00	1.03	113.3	142.0	137.1
Field pea	Hubal	2.66	2.26	2.54	160.3	207.7	219.5
	Batuta	2.85	2.16	2.21	171.6	219.1	226.3
Faba bean	Amulet	3.13	1.22	1.60	326.4	405.1	41.1
	Granit	3.19	1.32	1.62	361.5	447.7	415.6
$HSD_{0.05}$	For: species Cultivar	0.187 n.i.	0.213 n.i.	0.175 n.i.	17.8 10.6	18.8 n.i.	16.3 n.i.

Table 2.
Seed yield and weight of 1000 seeds of legume.

by pea cultivation, in particular the Hubal variety with the traditional form of foliage. This species significantly improved yields, especially in 2017 and 2018, with less favorable weather conditions during the growing season. Pea is a species with a shorter growing season, and earlier it started flowering and tying pods when the soil moisture was higher. On the other hand, the smallest level of yield was characterized by yellow lupine irrespective of the morphological type. The self-completing varieties of faba bean, yellow lupine, and blue lupine yielded at a higher level than varieties with a traditional type of growth. Among the pea varieties evaluated, the variety Hubal yielded the traditional type of foliage (**Table 2**).

Księżak [18] observed a higher level of yielding of the Ramrod pea variety (with bipinnate leaves) compared to the Rola variety (traditional). It was the result of this variety producing a longer fruiting stem, a greater number of pods, and a weight of seeds on the plant, as well as a weight of thousand seeds. Prusiński [19] and Szwejkowska et al. [20] comparing varieties of pea with diversified morphological structure noted better yielding of varieties with normal foliage. According to Podleśny and Strobel [21] and Bieniaszewski [22] from comparable varieties of blue lupine, they yielded better with traditional growth type than self-completing. Jarecki and Bobrecka-Jamro [23] recorded a higher yield of seeds of the Mister variety of yellow lupine than the self-completing ones. It resulted from the larger number of pods set up on the plant and the greater weight of 1000 seeds. Borowska and et al. [24] showed much higher yields of the traditional variety of white and yellow lupine than traditional ones. The same authors noted the opposite tendency in blue lupine. Also, according to Szymańska et al. [25], the Mister (traditional) cultivar yielded better on average in 5 years than the self-completing Perkoz. Podleśny [26], Prusiński [27], and Kulig [28] report that from among many evaluated faba bean varieties, Nadwiślański yielded much better than the self-completing ones. In other studies, Kulig [29] noted a more accurate yield of Kodam cultivar than Nadwiślański and Titus.

Significantly, higher protein yields as well as seed yields were noted in 2016 with favorable weather conditions during the growing season than in 2017 with a small amount of precipitation in June and the first decade of July. Significantly, higher yield of protein enabled the cultivation of faba bean, while the lower level of cultivation of pea and yellow lupine (**Table 3**). Obtained results by Panasiewicz et al. [30] indicate that among the four evaluated species (yellow lupine, white

Species	Cultivar	Protein content ($g{\cdot}kg^{-1}s.m$)		Protein yield ($kg{\cdot}ha^{-1}$)	
		2016	2017	2016	2017
Blue lupine	Kurant	300.5	312.5	759	440
	Regent	281.3	301.1	744	403
Yellow lupine	Bursztyn	425.2	437.3	658	371
	Perkoz	381.6	400.9	632	400
Field pea	Hubal	213.2	225.4	566	508
	Batuta	204.4	213.6	581	460
Faba bean	Amulet	287.6	306.8	898	373
	Granit	273.8	290.7	870	348
$HSD_{0.05}$	For: species Cultivar	29.6 14.28	31.70 1.29	17.7 12.4	20.7 2.5

Table 3.
The protein content and protein yield of legume.

lupine, blue lupine, and field pea), the highest yield of seeds and protein enables the cultivation of yellow lupine and the smallest of pea.

During the experimental period, changes in the structure of legume plants were observed depending on the course of the weather conditions during the growing season. Species and varieties were characterized by a varied plant structure. In 2018, all species established more pods and seeds on the plant, produced a large weight on the plant, and bound more seeds in the pod, and the blue lupine was characterized by a greater weight of 1000 seeds (**Table 4**). Peas among the assessed species were characterized by the smallest size of seeds; they formed the least pods and seeds on the plant and produced the smallest weight of seeds in the plant (**Table 5**).

Faba bean cultivars were characterized by a similar plant structure; only the Granit variety produced larger seeds and a larger number of seeds per plant. The self-completing variety of blue lupine compared to the traditional variety was characterized by a larger size of seeds, the number of pods, and weight of seeds on the plant, and the yellow lupine variety Perkoz greater number of pods and seeds on the plant and seeds in the pod (**Table 5**). In contrast, the pea variety Batuta (with bipinnate leaves) was characterized by a higher weight of thousand seeds, a greater number of pods, and weight of seeds on the plant. According to Borowiecki et al. [31], pea variety Wiato (with bipinnate leaves) was distinguished by the longer fruiting part, the greater number of pods per plant, and the greater weight of 1000 seeds compared to the traditional Rola variety. On the other hand, Podleśny and Podleśna [32] in the determined Legat variety of yellow lupine noted larger seeds than in the traditional Polo variety, and at the same time, the variety established more pods and seeds on the plant than the self-completing variety. Panasiewicz et al. [33] state that the traditional blue lupine variety Bojar was characterized by a greater number of pods and seeds per plant, a weight of thousand seeds, and Regent varieties produced more seeds in the pod. Szymańska et al. [25] observed a greater number of pods and seeds on the plant in the Mister variety of yellow lupine than in the Perkoz variety (self-completing), and the number of seeds in the pod and the weight of thousand seeds were similar in both varieties. Podleśny [26] states that the self-completing variety of faba bean Tim planted more pods on the plant, and the Nadwiślański variety set more seeds on the plant and was characterized by larger seeds. According to Prusiński [27], the

Species	Cultivar	Number of pods			Number of seeds		
		2016	2017	2018	2016	2017	2018
Blue lupine	Kurant	7.68	6.18	7.90	33.1	23.9	28.36
	Regent	8.15	4.88	8.35	37.6	19.2	36.25
Yellow lupine	Bursztyn	7.73	4.35	8.60	28.6	14.5	28.29
	Perkoz	8.6	5.46	8.45	31.1	16.2	26.70
Field pea	Hubal	5.15	3.05	4.60	18.8	9.8	17.66
	Batuta	5.45	2.65	4.50	20.6	8.7	16.38
Faba bean	Amulet	11.08	2.90	5.35	35.6	7.5	15.62
	Granit	10.05	3.40	5.33	34.9	8.3	15.83
$HSD_{0.05}$	For: species Cultivar	0.314 n.i.	0.058 n.i.	0.240 n.i.	1.86 1.32	1.72 0.06	1.637 n.i.

Table 4.
Number of pods and seeds on plant.

Species	Cultivar	Weight of seeds on plant (g)			Number of seeds per pods		
		2016	2017	2018	2016	2017	2018
Blue lupine	Kurant	4.63	3.13	4.18	4.30	3.87	3.59
	Regent	5.20	2.72	4.77	4.65	3.93	4.34
Yellow lupine	Bursztyn	2.98	1.89	3.38	3.73	3.33	3.29
	Perkoz	3.60	2.30	3.71	3.60	2.97	3.16
Field pea	Hubal	3.08	2.17	3.32	3.53	3.21	3.84
	Batuta	3.45	1.78	3.04	3.70	3.30	3.64
Faba bean	Amulet	11.58	3.03	5.42	3.23	2.59	2.92
	Granit	10.20	3.70	5.33	3.48	2.44	2.97
$HSD_{0.05}$	For: species Cultivar	0.216 0.018	0.132 n.i.	0.079 0.118	0.170 0.149	0.152 0.089	0.077 0.075

Table 5.
The weight of seeds on plant and number of seeds per pods.

Nadwiślański variety was characterized by more favorable elements of the yield structure (number of pods, seeds, seed mass, weight of thousand seeds) than the self-completing varieties.

The longest fruiting part was produced by faba bean, especially the Amulet variety with a traditional type of growth. In the other varieties of the assessed species, the values of these features were relatively small. In 2016 and 2017, the plants of the evaluated species, the first pod, were established at a similar height; only in 2018 both varieties of peas deposited it much higher (**Tables 6–8**). Both pea varieties were characterized by the smallest dry mass of stems and pods than those of other species (**Table 9**). Kulig [29] observed the highest faba bean plants of the traditional type of growth (Nadwiślański), and the Titus (self-completing) variety deposited only the first pod and produced the shortest fruiting part.

Fat is an important component of legume seeds, regardless of the species. The seeds of yellow and blue lupine contained more than seeds of pea and faba bean. Varieties of peas accumulated a similar amount of this ingredient, while more fat

Species	Cultivar	Height of first pod	Height of last pod	Height of the apex pea plant	Length of fruiting pods
Blue lupine	Kurant	42.8	51.2	55.6	8.4
	Regent	41.3	52.3	58.2	11.0
Yellow lupine	Bursztyn	50.9	56.1	64.3	5.2
	Perkoz	46.3	58.5	66.6	12.2
Field pea	Hubal	50.0	58.2	64.9	8.2
	Batuta	49.3	59.0	68.6	9.7
Faba bean	Amulet	45.2	68.1	93.5	22.9
	Granit	45.0	69.8	82.8	24.8

Table 6.
The height of the first pod, of the last pod, and of the apex pea plant and length of the fruiting part of the stem in 2016.

Species	Cultivar	Height of the first pod	Height of the last pod	Height of the apex pea plant	Length of the fruiting part of the stem
Blue lupine	Kurant	31.8	38.4	41.5	6.6
	Regent	27.2	32.3	37.8	5.1
Yellow lupine	Bursztyn	37.2	44.5	54.1	7.3
	Perkoz	34.5	40.9	52.4	6.4
Field pea	Hubal	43.9	47.5	52.7	3.6
	Batuta	44.5	48.2	52.4	3.7
Faba bean	Amulet	36.7	48.3	63.1	11.6
	Granit	36.8	46.7	60.1	9.9

Table 7.
The height of the first pod, of the last pod, and of the apex pea plant and length of the fruiting part of the stem in 2017.

contented in self-completing varieties of faba bean, blue lupine, and yellow lupine (**Table 10**). The studies showed that, regardless of the agroecological conditions, the yellow lupine varieties were characterized by a higher content of crude protein and crude fiber. Both blue lupine varieties contents high amount of fiber also. Unfavorable weather conditions during the growing season had a positive effect on the accumulation of protein and fat in the seeds of all legume species.

Podleśny and Strobel [21], among the evaluated varieties of blue lupine, the most favorable protein content was characterized by Graf varieties, yellow lupine Wersal, and the most fiber accumulated in varieties of Graf and Boruta. The same authors [21] state that the amount of ash in seeds of all varieties was similar. Podleśny and Strobel [34] did not report differences in protein concentration in seeds of yellow lupine varieties. In their opinion, more fat contained Legat and Markiz seeds, and the least fiber of Polo variety. In their studies, ash content was similar in all species and varieties. Kulig [28] did not observe differences in protein content in the seeds of various faba bean cultivars with different morphological structures.

Księżak et al. [35] in previous studies showed that regardless of the habitat conditions varieties of blue lupine (Graf and Tango) and yellow lupine (Dukat,

Species	Cultivar	Height of the first pod	Height of the last pod	Height of the apex pea plant	Length of the fruiting part of the stem
Blue lupine	Kurant	54.6	53.6	59.3	5.7
	Regent	40.8	49.1	55.1	6.0
Yellow lupine	Bursztyn	43.2	60.1	65.2	5.1
	Perkoz	37.9	47.2	52.2	5.5
Field pea	Hubal	58.4	65.6	69.8	4.2
	Batuta	59.0	65.6	70.3	3.7
Faba bean	Amulet	41.6	61.6	76.7	15.1
	Granit	45.6	53.2	63.0	9.8

Table 8.
The height of the first pod, of the last pod, and of the apex pea plant and length of the fruiting part of the stem in 2018.

Species	Cultivar	Dry matter of the stem of one plant			Dry matter of siliques		
		2016	2017	2018	2016	2017	2018
Blue lupine	Kurant	4.25	1.60	4.48	2.28	1.54	2.16
	Regent	4.80	1.59	3.40	2.30	1.44	2.08
Yellow lupine	Bursztyn	4.05	1.92	5.25	2.80	1.92	3.58
	Perkoz	4.30	2.09	4.99	2.64	1.67	3.00
Field pea	Hubal	2.73	2.19	3.77	0.75	0.39	0.62
	Batuta	2.80	1.96	4.09	0.78	0.32	0.69
Faba bean	Amulet	11.20	3.34	5.59	3.15	1.63	1.96
	Granit	8.63	3.23	5.23	3.05	1.72	2.18
$HSD_{0.05}$	For: species Cultivar	0.227 0.283	0.184 0.012	0.090 0.096	0.148 n.i.	0.193 0.019	0.088 0.068

Table 9.
Dry matter of the stem of one legume plant and siliques (g).

Talar, Lord, and Baryt) were characterized by the highest protein content, while the Sonet variety of blue lupine and the Perkoz variety of yellow lupine—the smallest. Lagunes-Espinoza et al. [36] inform that the protein content in seeds in the same lupine species is relatively little differentiated, while definitely larger differences occur between species. Rybiński et al. [37] also report that among the varieties of blue lupine the largest amount of protein was recorded in the seeds of Graf, Baron, Neptun, and Boruta. Niwińska [38] reports that much less proteins contain sweet lupine seeds than alkaloid ones.

Obtained results by Księżak et al. [35] indicate that the evaluated varieties of yellow lupine were characterized by a similar fiber content; only the Perkoz variety contained significantly more of this component than Parys, whereas in the case of blue lupine, it contained the least Neptun variety and indeed more varieties Karo, Boruta, Graf, Bojar, and Kadryl. Niwińska [39] noted species and varietal diversity in fiber accumulation. The most of this ingredient contained the blue lupine variety Sur, the least white lupine Bardo variety, and blue lupine Emir variety. The authors mentioned earlier [35] state that the least fat from the included varieties

Species	Cultivar	Ash		Fat		Fiber	
		2016	2017	2016	2017	2016	2017
Blue lupine	Kurant	42.3	43.4	62.4	67.2	173.1	162.1
	Regent	33.9	36.7	64.2	71.2	162.4	163.5
Yellow lupine	Bursztyn	40.1	42.6	50.4	60.2	172.3	158.2
	Perkoz	38.4	39.3	59.5	73.1	176.4	157.6
Field pea	Hubal	33.2	33.4	28.0	32.6	69.1	64.2
	Batuta	35.6	36.3	28.8	34.3	66.8	61.3
Faba bean	Amulet	37.1	36.4	21.9	24.0	91.1	94.3
	Granit	36.8	37.1	30.1	28.3	87.2	90.3
$HSD_{0.05}$	For: species cultivar	3.74 n.i.	0.701 0.62	2.41 3.10	1.43 1.24	9.55 n.i.	8.71 n.i.

Table 10.
Concentrations of crude fiber and fat in faba bean seeds depending on the method of fertilization (%).

of blue lupine was collected by Boruta, while significantly more varieties were Kalif, Regent, and Zeus. On the other hand, in the case of yellow lupine, the Perkoz variety was distinguished by a significantly higher content of this ingredient than the other varieties.

Obtained by Księżak et al. [40], the results of the protein content assessment in faba bean showed that the smaller amount of it characterized Sonet, Optimal, and Granit varieties, while the larger the other evaluated varieties. Sarah et al. [41] inform that the content of protein, carbohydrates, ash, fat, and fiber depends on the variety. Mekkei [42], that regardless of the varieties, large bean seeds contain more protein and carbohydrates. Hendawey [43] reports that the differences in features between faba bean cultivars are caused by both genetic and environmental factors. Księżak [44] similar content of protein, fiber, fat ash, and nitrogen-free extract compounds in reported Nadwiślański, Bronto, Tino, and Martin varieties. Only Caspar varieties contained less protein and more nitrogenous compounds [44].

Species	Cultivar	Potassium		Phosphorus	
		2016	2017	2016	2017
Blue lupine	Kurant	10.2	10.2	4.20	4.42
	Regent	10.3	9.7	4.23	4.19
Yellow lupine	Bursztyn	12.3	11.4	5.12	5.21
	Perkoz	11.7	11.2	5.02	5.09
Field pea	Hubal	11.3	10.8	4.93	4.83
	Batuta	10.8	11.1	5.28	5.30
Faba bean	Amulet	12.5	11.6	6.71	6.82
	Granit	11.6	12.0	6.82	6.91
$HSD_{0.05}$	For: species Cultivar	1.58 0.62	1.63 n.i.	0.07 0.06	0.08 n.i.

Table 11.
Concentrations of potassium and phosphorus in seeds depending on legume cultivar (%).

Księżak [40] reported varied content of fiber, the least contained it Bobas and Granitp. Księżak [40] noted similar fat content in all evaluated cultivars in faba bean. However, Hendawey [43] showed greater concentration in Giza 843 and Giza 3. Nowacka-Zaborska and Oleszek [45] observed higher content of fat in faba bean seeds in drought conditions. The obtained results indicate that the seeds of both faba bean species showed a higher concentration of phosphorus and potassium in comparison with other species (statistically significant differences) (**Table 11**). There were no significant differences between the compared varieties within all plant species.

4. Conclusions

In ecological conditions, pea cultivation especially the Hubal variety (traditional form of foliage) allowed to obtain the largest seed yield and the smallest cultivation of yellow lupine independent on the morphological type. The self-completing varieties of faba beans, blue lupines, and yellow lupines were yielded at a higher level than varieties with a traditional type of growth. Significantly, higher yield of protein is provided by faba bean cultivation, while the smaller level of pea and yellow lupine.

Yellow and blue lupine seeds contained more fat than pea and faba bean seeds. Pea varieties, regardless of the form of foliage, accumulated a similar amount of this component, while more self-completing varieties of faba bean, blue lupine, and yellow lupine. Irrespective of the agroecological conditions, the seeds of the yellow lupine varieties were characterized by a higher protein and fiber content. Both varieties of blue lupine also characterized high fiber content. Unfavorable weather conditions during the growing season have positively influenced the accumulation of protein and fat in the seeds of all legume species. The seeds of the tested species contained a similar amount of potassium and phosphorus, a greater amount of ash characterized blue lupine of the Kurant variety.

Author details

Księżak Jerzy and Bojarszczuk Jolanta*
Department of Forage Crop Production, Institute of Soil Science and Plant Cultivation—State Research Institute, Puławy, Poland

*Address all correspondence to: jbojarszczuk@iung.pulawy.pl

References

[1] Jansen G. Effects of temperature on yield and protein content of *Lupinus angustifolius* cultivars. In: Palta JA, Berger JB, editors. Lupins for Health and Wealth. Proceedings of the 12th International Lupin Association, Canterbury, New Zealand; 2008. pp. 342-345

[2] Prusiński J. The legumes in European Union. Zeszyty Problemowe Postepów Nauk Rolniczych. 2010;**550**:11-19

[3] UN. General Assembly. Resolution adopted by the General Assembly on 20 December 2013 (on the report of Second Committee (A68/444). 68/231. International Year of Pulses; 2016

[4] Vioque J, Alaiz M, Girón-Calle J. Nutritional and functional properties of *Vicia faba* protein isolates and related fractions. Food Chemistry. 2012;**132**:67-72. DOI: 10.1016/j.foodchem.2011.10.033

[5] Summo C, Centomani I, Paradiso VM, Caponio F, Pasqualone A. The effects of the type of cereal on the chemical and textural properties and on the consumer acceptance of pre-cooked, legume-based burgers. LWT–Food Science and Technology. 2016;**65**:290-296

[6] Campos-Vega R, Loarca-Pina G, Oomah BD. Minor components of pulses and their potential impact on human health. Food Research International. 2010;**43**(2):461-482. DOI: 10.1016/j.foodres.2009.09.004

[7] Ramos S. Effects of dietary flavonoids on apoptotic related to cancer chemoprevention. The Journal of Nutritional Biochemistry. 2007;**18**:427-442

[8] Lqari H, Vioque J, Pedroche J, Millan F. *Lupinus angustifolius* protein isolates: Chemical composition, functional properties and protein characterization. Food Chemistry. 2002;**76**:349

[9] Mateos-Aparicio I, Redondo-Cuenca A, Villanueva-Suarez M, Zapata-Revilla M, Tenorio-Sanz M. Pea pod, broad bean pod and okara, potential sources of functional compounds. LWT–Food Science and Technology. 2010;**43**(9):1467-1470. DOI: 10.1016/j.lwt.2010.05.008

[10] Duranti M, Consonni A, Magni C, Sessa F, Scarafoni A. The major proteins of lupin seed: Characterization and molecular properties for use as functional and nutraceutical ingredients. Food Science and Technology. 2008;**19**:624

[11] Small E. Lupines—Benefit and harm potentials. Biodiversity. 2012;**13**:54

[12] Doležal P, Zeman L, Skládanka J. Effect of supplementation of chemical preservatives on fermentation process of lupine silage. Slovak Journal of Animal Science. 2008;**1**:30-38

[13] Fraszer MD, Fychan R, Jones R. Comparative yield and chemical composition of two varieties of narrow-leafed lupin (*Lupinus angustifolius*) when harvested as whole-crop, moist grain and dry grain. Animal Feed Science and Technology. 2005;**120**:43-50

[14] Szyszkowska A, Bodarski R, Sowiński J, Załęska A. The possibilities of using green forages from intercroppings of maize and field bean as the raw material for silage. Zeszyty Problemowe Postepów Nauk Rolniczych. 2007;**522**:361-370

[15] Podleśna A, Podleśny J, Doroszewski A. Usefulness of selected weather indices to evaluation of yellow lupine yielding possibility. Agricultural Water Management. 2014;**146**:201-207

[16] Faligowska A, Panasiewicz K, Szymańska G, Szukała J, Koziara W, Pszczółkowska A. Productivity of white lupin (*Lupinus albus* L.) as an effect of diversified farming systems. Legume Research. 2017;**40**:872-877

[17] Atkins CA, Smith PM. Regulation of pod set and seed development in lupin. In: van Santen E, Hill GD, editors. In: Proceedings of the 10th International Lupin Conference, 19-24. June 2002, New Zealand; 2004. pp. 275-278

[18] Księżak J. Influence of selected herbicides on development and yielding of semileafless variety of pea. Progress in Plant Protection. 2007;**3**:169-172

[19] Prusiński J. Chosen growth and development indexes of pea under increasing intensity of cultivation technology. Acta Scientiarum Polonorum, Agricultura. 2007;**6**(4):43-51

[20] Szwejkowska B. Reaction of pea (*Pisum sativum* L.) cultivars to different weed control methods. Acta Scientiarum Polonorum, Agricultura. 2006;**5**(1):71-82

[21] Podleśny J, Strobel W. The effect of sowing date on seed and protein yield formation of differentiated genotypes of blue lupine. Acta Agrophysica. 2006;**8**(4):923-933

[22] Bieniaszewski T, Podleśny J, Olszewski J, Stanek M, Kaszuba M. The response of indeterminate and determinate narrow-leaved lupin varieties to different plant density. Fragmenta Agronomica. 2012;**29**(4):21-35

[23] Jarecki W, Bobrecka-Jamro D. Effect of the sowing date on the size and quality of the seed yield of yellow lupine (*Lupinus luteus* L.). Acta Scientiarum Polonorum, Agricultura. 2014;**13**(2):13-22

[24] Borowska M, Prusinski J, Kaszkowiak E. Production results of intensification of cultivation technologies in three lupin (Lupinus L.) species. Plant, Soil and Environment. 2015;**61**:426-431

[25] Szymańska G, Faligowska A, Panasiewicz K, Szukała J, Koziara W. The productivity of two yellow lupine (*Lupinus luteus* L.) cultivars as an effect of different farming systems. Plant, Soil and Environment. 2017;**63**(12):552-557. DOI: 10.17221/639/2017-PSE

[26] Podleśny J. Effect of amount and distribution of precipitation during vegetation on growth, development and yielding of determinate and traditional faba bean varieties. Acta Agrophysica. 2009;**14**(2):413-425

[27] Prusiński J. Effect of plant density on self-completing faba bean (*Vicia faba* ssp. minor) cultivars yielding on light soil. Acta Scientirum Polonorum Agricultura. 2003;**2**(2):107-118

[28] Kulig B, Oleksy A, Sajdak A. Yielding of selected faba bean cultivars depending on plant protection methods and sowing density. Fragmenta Agronomica. 2009;**26**(3):93-101

[29] Kulig B, Pisulewska E, Sajdak A. Effect of sowing rate on the yield and size of assimilation area in selected field bean cultivars. Zeszyty Problemowe Postępów Nauk Rolniczych. 2007;**522**:263-270

[30] Panasiewicz K, Koziara W, Sulewska H, Szukała J, Faligowska A, Szymańska G, Ratajczak K, et al. Productivity of selected species of fabaceae in reduced tillage conditions within a production field. Nauka Przyroda Technologie 2016;**10**(1). ISSN 1897-7820. DOI: 10.17306/J.NPT.2016.1.6

[31] Borowiecki J, Księżak J, Bournoville R, Lerin J. Influence of weevil beetle

feed on the development and yields of pea. Pamiętnik Puławski. 2004;**137**:5-14

[32] Podleśny J, Podleśna A. The effect of high temperature in the period of flowering on growth, development and yielding of yellow lupine. Acta Agrophysica. 2012;**19**(4):825-834

[33] Panasiewicz K, Faligowska A, Szymańska G, Koziara W, Szukała J, Poniatowska J. Yielding of narrow-leaved lupin depending on varieties, sowing method and sowing rate. Fragmenta Agronomica. 2018;**35**(1): 72-80. DOI: 10.26374/fa.2018.35.07

[34] Podleśny J, Strobel W. The effect of variety and sowing date on yield and amino-acid composition of yellow lupine seed protein. Acta Agrophysica. 2007;**10**(1):175-185

[35] Księżak J, Staniak M, Bojarszczuk J. Nutrient contents in yellow lupine (*Lupinus luteus* L.) and blue lupine (*Lupinus angustifolius* L.) cultivars depending on the habitat conditions. Polish Journal of Environmental Studies. 2018;**27**(3):1-9. DOI: 10.15244/pojes/76677

[36] Lagunes-Espinoza LC, López-Upton J, García-López E, Jasso-Mata J, Delgado-Alvarado A, García de Los Santos G. Morphological diversity and protein concentration of Lupinus spp. in central-eastern region of the state of Puebla. Acta Botanica Mexicana. Pátzcuaro abr. 2012. p. 99

[37] Rybiński W, Starzycki IM, Rusinek R, Bocianowski J, Szot B. Variability of chemical composition of legume seeds and their resistance to mechanical loads. Biuletyn IHAR. 2013;**268**:93-209

[38] Niwińska B. Composition and nutritive value of crude and prepared proteins of various lupine varieties. Roczniki Naukowe Zootechniki. 1996;**23**(3):229-237

[39] Niwińska B. The nutritive value of Polish-grown lupin cultivar seeds for ruminants. Journal of Animal and Feed Sciences. 2001;**10**:91-101

[40] Księżak J, Bojarszczuk J, Staniak M. Evaluation of the Concentration of Nutrients in the Seeds of Faba Bean (*Vicia faba* L. major) and Pea (*Pisum sativum* L.) Depending on Habitat Conditions. Polish Journal of Environmental Studies. 2018;**27**(3):1-11. DOI: 10.15244/pjoes/76175

[41] Sarah A, Abusin E, Hassan AB, Babiker EE. Nutritional evaluation of cooked faba bean (*Vicia faba* L.) and white bean (*Phaseolus vulgaris* L.) cultivars. Australian Journal of Basic and Applied Sciences. 2009;**3**(3):2484-2490

[42] Mekkei ME. Effect of intra-row spacing and seed size on yield and seed quality of faba bean (*Vicia faba* L.). International Journal of Agriculture and Crop Sciences. Available online: www.ijagcs.com. IJACS/2014/7-10/665-670. ISSN 2227-670X ©2014

[43] Hendawey MH, Younes AMA. Biochemical evaluation of some faba bean cultivars under rainfed conditions at El-Sheikh Zuwayid. Annals Agriculture Science. 2013;**58**(92):183-193

[44] Księżak J. Dynamics of nutrient uptake by traditional and self-finishing varieties of horse bean between flowering and full maturity. Monografie i Rozprawy Naukowe. 2002;**5**:1-95

[45] Nowacka-Zaborska J, Oleszek W. The concentration of oligosaccharide and fan in the seeds of polish faba bean (*Vicia faba*) cultivars. Pamiętnik Puławski. 1995;**106**:139

Chapter 7

Influence of Adjuvants on Efficacy of Postemergence Herbicides Commonly Used in Peanut (*Arachis hypogaea* L.)

William James Grichar, Peter A. Dotray and Mark A. Matocha

Abstract

Field studies were conducted for 2 years in the High Plains of Texas (34.1826° N, 101.9505° W) and in South Texas (29.1634° N, 97.0725° W) to evaluate weed control when using different adjuvants with commonly used peanut herbicides. In the High Plains, *Amaranthus palmeri* L. control with acifluorfen, imazapic, lactofen, and 2,4-DB at the 1X dose improved with the use of an adjuvant over no adjuvant. *A. palmeri* control with imazethapyr was similar to that seen with imazapic and lactofen with the exception of the 1/2X rate of imazethapyr, which showed improved control with Agridex over the use of no adjuvant or Induce in 1 year, while Induce was better than no adjuvant or Agridex in the other year. In 1 year in South Texas, *A. palmeri* control with imazapic at the 1X dose was ≥73% with/without an adjuvant. In another year, the 1X dose of imazapic controlled *A. palmeri* 64% without an adjuvant, while the addition of Cide Kick II resulted in 83% control. An adjuvant did not improve *A. palmeri* control with lactofen or *Cucumis melo* L. control with either imazapic or lactofen. *Urochloa texana* (Buckl.) control with clethodim at the 1X dose was not improved by the addition of an adjuvant in either year. *U. texana* control was not improved when using the 1X dose of fluazifop-P with any adjuvant.

Keywords: herbicides, Palmer amaranth, smell melon, Texas millet, weed control

1. Introduction

An adjuvant is described as any compound that lowers the surface tension of a liquid, thereby increasing the contact between the liquid and another substance [1]. The efficacy of postemergence (POST) herbicides is influenced by several factors including weed species [2], weed size [2, 3], environmental conditions at the time of application [4, 5], application rate [2], interactions with other agrichemicals [6, 7], and the interaction with adjuvants [3, 7–10].

Adjuvants enhance herbicide efficacy primarily through increasing herbicide absorption [9–12]. Some adjuvants alter the formulation of a herbicide so that the herbicide more completely and evenly covers the plant surfaces, thereby keeping the herbicide in contact with plant tissue rather than beading up and rolling off [13, 14]. This is accomplished by the adjuvant reducing the surface tension and contact angle of herbicide solution, thereby improving the coverage of the solution and improving the chance for the herbicide to penetrate the plant surface [15–17].

Foy and Smith [18] studied the effect of adjuvants on surface tension and herbicide efficacy and found that minimum surface tension and contact angle occurred at concentrations of 0.1–0.5% for all adjuvants tested. However, maximum herbicidal activity was observed at 1% concentration, which indicated that there were other factors increasing herbicide activity besides surface tension and contact angle. They concluded that specific interactions of herbicide-adjuvant-plant surface were a part of the total adjuvant action.

Other adjuvants increase the herbicides' penetration through the cuticular wax, cell walls, and/or stomatal openings [13, 14, 19, 20]. Crop oil concentrates and vegetable oils fall into the category of penetrants [20]. This type of adjuvant improves cuticular penetration by softening, plasticizing, or dissolving cuticular waxes and allowing herbicide movement to the more hydrophilic regions underneath [20]. Although volatile herbicides easily penetrate stomata, stomatal penetration by an aqueous solution is not possible unless the surface tension of the spray solution is reduced significantly [20]. Most adjuvants are incapable of reducing surface tension enough to allow stomatal penetration. Prior to the development of the organosilicone surfactants, stomatal infiltration of herbicides into the leaf was considered to be of minor importance [20]. In contrast to other wetting agents, the organosilicone surfactants can reduce surface tension to levels low enough to allow stomatal infiltration of aqueous spray solutions [21–23]. When stomatal penetrations occur, it is greatest in the morning when stomates are more likely to be open.

The objectives of this research were (1) to compare efficacy of several grass and broadleaf herbicides commonly used in peanut (*Arachis hypogaea* L.) when applied with different adjuvants and (2) compare the different spray adjuvants when labeled and sublethal herbicide doses are used with acifluorfen, clethodim, fluazifop-P-butyl, imazapic, imazethapyr, lactofen, and 2,4-DB on four major weeds found in Texas peanut.

2. Materials and methods

2.1 Field studies

These studies were conducted during the 2011 and 2012 peanut growing seasons in the Texas High Plains near Halfway (34.1826° N, 101.9505° W) and during the 2012 and 2013 growing seasons in the south-central Texas peanut growing region near Yoakum (29.276° N, 97.123° W). Soil type at the High Plains location was a Pullman clay loam (fine, mixed, thermic Torrertic Paleustoll) with less than 1% organic matter and pH 7.7, while at the South Texas location, the soil type was a Denhawken sandy loam (fine-silty, carbonitic, hyperthermic Fluventic Ustochrepts) with less than 1.0% organic matter and pH 7.6. Studies were conducted in the same field but moved from year-to-year to different areas within those fields. Irrigation was applied as needed to maintain soil moisture and plant growth.

2.2 Herbicides, doses, and application

Postemergence herbicide treatments at the High Plains location included acifluorfen {5-[2-chloro-4-(trifluoromethyl) phenoxy]-2-nitrobenzoic acid} at 0.28 (1/2X) and 0.56 kg ai/ha (1X), imazapic {(+)-2-[4,5-dihydro-4-methyl-4-4(1-methylethyl)-5-oxo-1H-imidazol-2-yl]-5-methyl-3-pyridinecarboxylic acid} at 0.035 (1/2X) and 0.07 kg ai/ha (1X), imazethapyr {2-[4,5-dihydro-4-methyl-4-(1-methylethyl)-5-oxo-1H-imidazol-2-yl]-5-ethyl-3-pyridinecarboxylic acid}

Adjuvant	Adjuvant composition	Dose (%, v/v)	Manufacturer
Agridex	Paraffin-based petroleum oil (83%) and surfactant blend (17%)	1.0	Helena Chem. Co.
Cide-Kick II	100% d'limonene and related isomers plus selected emulsifiers	1.0	Brewer International
ETA	Paraffinic petroleum oil (60%) and ethoxylated nonionic surfactant (40%); unsulfonated oil residue (UR) value, 90% minimum	1.0	Aurora Cooperative
Induce	Alkylarylpolyoxylkane ether, free fatty acids isopropyl (90%) and water and formulation aids (10%)	0.25	Helena Chem. Co.
90–10	Alkyl, polyethoxy ethers, ethoxylated and soybean derivatives, and antifome 90–10	1.0	Precision Laboratories

Table 1.
Adjuvants, composition, dose, and manufacturer.

at 0.035 (1/2X) and 0.07 kg ai/ha (1X), lactofen {2-ethoxyl-1-methyl-2-oxoethyl 5-[2-chloro-4-(trifluoromethyl)phenoxy]-2-nitrobenzene} at 0.11 (1/2X) and 0.22 (1X) kg ai/ha, and 2,4-DB [4-(2,4-dichlorophenoxy) butanoic acid] at 0.14 (1/2X) and 0.28 (1X) kg ai/ha. An untreated check was included for comparison.

In South Texas, herbicides in the broadleaf weed study included imazapic and lactofen at the previously mentioned rates, while in the annual grass study, the herbicides included clethodim {(E)-2-[1-[[(3-chloro-2-propenyl) oxy]imino]propyl]5-[2-(ethylthio)propyl]-3-hydroxy-2-cyclohexen-1-one} at 0.05 (1/2X) and 0.1 (1X) kg ai/ha and fluazifop-P-butyl {(butyl)(R)-2-[4-[[5-(trifluoromethyl)-2-pyridinyl]oxy]phenoxy]propanoate} at 0.1 (1/2X) and 0.2 (1X) kg ai/ha. Adjuvants in both studies include Agridex®, Cide-Kick II®, ETA®, Induce®, and 90–10® (**Table 1**). An untreated check was included for comparison in each study.

Herbicides were applied in water using a CO_2 pressurized backpack sprayer with TeeJet® 11002 DG (Spraying Systems Company, P.O. Box 7900, North Avenue, Wheaton, IL 60188) nozzles calibrated to deliver 190 L/ha at 180 kPa at the South Texas location and TurboTee® 110015 nozzles calibrated to deliver 94 L/ha at 207 kPa at the High Plains location. Herbicides were applied POST when *Amaranthus palmeri* L. was up to 43 cm tall, while *Cucumis melo* L. var. Dudaim Naud. vines were vining up to 38 cm in length. *Urochloa texana* (Buckl.) R. Webster was up to 46 cm in height at the time of herbicide application (**Table 2**).

2.3 Experimental design, weeds, and densities

The experimental design was a randomized complete block with three replications at both locations. *A. palmeri* was evaluated at the High Plains location in a 5-(herbicide)-by-2-(dose)-by-3 (adjuvant) factorial arrangement of treatments.

At the South Texas location, two separate studies were completed. In the first study, *A. palmeri* and *C. melo* were evaluated using imazapic and lactofen, while in another study, *U. texana* was evaluated using clethodim and fluazifop-P butyl for control. Both studies in South Texas were a 2-(herbicide)-by-2-(dose)-by-6-(adjuvant) factorial arrangement of treatments.

Individual plots at the High Plains location were four rows 9.5 m long spaced 101 cm apart, and the middle two rows of each plot were sprayed, while at the South Texas location, plots were two rows 7.9 m long spaced 97 cm apart. Natural

Date	Time of day	Herbicide appl.	Dew	RH (%)	AT (°C)	ST (°C)	SM[a]	WS[a, b] (cm)
High Plains								
6/30/2011	10:30 am	All	No	35	32	26	G	10–36
6/02/2012	2:30 pm	All	No	51	29	31	D	25–43
South Texas								
7/19/2012	7:45 am	Imazapic lactofen	Yes	90	31	27	E	15–36(A) 15–38(C)
7/23/2012	8:00 am	Clethodim fluazifop	Yes	98	25	27	E	20–46(U)
7/13/2013	6:30 am	Imazapic	Yes	90	24	27	G	15–25(A) 15–30(C)
7/14/2013	6:30 am	Lactofen	Yes	94	24	27	G	15–25(A) 15–30(C)
7/23/2013	7:00 am	Clethodim	Yes	96	26	26	E	20–46(U)
7/24/2013	7:00 am	Fluazifop	Yes	96	24	26	E	20–46(U)

[a]*A, A. palmeri; AT, air temperature; C, C. melo; D, dry; E, excellent; G, good; RH, relative humidity; ST, soil temperature; SM, soil moisture; U, U. texana; and WS, weed size.*
[b]*Only A. palmeri was present at Halfway.*

Table 2.
Environmental conditions at time of herbicide application at each location.

infestations of *A. palmeri* at the High Plains location were present at a population range of 6–8 plants/m^2. *A. palmeri*, *C. melo*, and *U. texana* were present in South Texas at a population density of 6–10 plants/m^2 in both years.

2.4 Peanuts and planting

At the High Plains location, OLin [24] was planted in both years at the rate of 100 kg/ha. Planting date in 2011 was April 27, while in 2012, the planting date was May 1. Tamrun OL07 [25] and Georgia 09B [26] peanut were planted at the rate of 110 kg/ha in South Texas on June 14, 2012 and June 6, 2013, respectively. At neither location was peanut harvested for yield.

2.5 Weed efficacy ratings and data analysis

Weed control or peanut injury was estimated visually using a scale of 0 (no weed control or peanut injury) to 100 (complete weed control or plant death) relative to the untreated control [27]. Weed control ratings and peanut injury consisting of chlorosis and/or stunting (where applicable) were taken 2 and 4 weeks after herbicide application.

Data from the High Plains were analyzed using a five by a two-by-three factorial analysis (POST herbicide by dose by adjuvant), while the data from South Texas were analyzed using a two-by-two-by-six factorial analysis (POST herbicides by dose by adjuvant). Significant differences among treatments were determined using analysis of variance and means were separated by protected Fisher's LSD test at $P < 0.05$ [28]. Visual estimates of weed control and peanut injury were transformed to the arcsine square root prior to analysis of variance, but are expressed in their original form for clarity because the transformation did not alter interpretation. The untreated check was not included in the weed control or peanut injury analysis.

3. Results and discussion

3.1 *Amaranthus palmeri* control

3.1.1 High Plains of Texas

No attempt was made to consolidate data over years since there was a treatment by year interaction and environmental conditions (relative humidity, soil temperature, and soil moisture) at time of herbicide application varied between years (**Table 2**). Also, extremely hot, dry weather conditions were observed in 2011 (data not shown). Although the test area was irrigated, the record high temperatures and low rainfall [29] made it difficult to maintain adequate soil moisture for plant growth.

In 2011, only the high dose of acifluorfen and 2,4-DB showed no response to the addition of an adjuvant, while the addition of either Agridex or Induce to the low dose of acifluorfen and 2,4-DB improved *A. palmeri* control over those herbicides with no adjuvant (**Table 3**). The addition of Induce to either imazapic or imazethapyr at 0.035 kg/ha or lactofen at 0.11 kg/ha improved control over those herbicides without any adjuvant, while the addition of Agridex to the high dose of these herbicides improved control over Induce or the use of the herbicide with no adjuvant. Other research has reported that herbicide rates can be reduced up to 75% with the use of adjuvants, usually when applications are made during early growth stages [30–32]. However, successful control using reduced herbicide rates depends on weed growth stage sensitivity [33, 34] and current environmental conditions [35, 36].

In 2012, the low dose of either imazapic or lactofen showed no response to *A. palmeri* control with the addition of an adjuvant, while acifluorfen, imazapic, imazethapyr, or lactofen at the high dose and 2,4-DB at both doses resulted in greater control with the addition of either Agridex or Induce over the use of no adjuvant (**Table 3**). *A. palmeri* control with acifluorfen, imazapic, or lactofen herbicides was similar with either adjuvant. Imazethapyr, at either dose, provided better control with the addition of Agridex than the addition of Induce. Since soil moisture was low in 2011 and weed size at time of herbicide application was greater in 2012 than 2011 (**Table 2**), the use of an adjuvant proved beneficial. Adjuvants have been reported to increase absorption of bentazon in *Abutilon theophrasti* Medic. [37] although plants were water-stressed [38]. Bellinder et al. [39] reported that there was no benefit in using a crop oil concentrate (COC) with either bentazon or fomesafen at the 0–2 or 2–4-leaf stage of *A. theophrasti*; however, control was inconsistent at the 4–6-leaf stage even when a COC was used.

3.1.2 South Texas

In 2012, only the addition of ETA to imazapic at the low dose improved *A. palmeri* control over the use of either imazapic or lactofen without an adjuvant (**Table 4**). In 2013, the addition of either Induce or Cide-Kick II to the low dose of imazapic or Cide-Kick II and 90–10 to the high dose of imazapic improved control over both doses of imazapic without an adjuvant. No other adjuvants improved *A. palmeri* control over either dose of imazapic or lactofen without an adjuvant.

In both years, *A. palmeri* amaranth control when using lactofen with or without an adjuvant was at least 88% with the exception of the addition of ETA to the high dose of lactofen in 2012, which resulted in 78% control. Grichar and Dotray [40] reported that lactofen control of *A. palmeri* was greater when applied to 2–5 cm tall compared with either 15–20 or 25–30 cm tall plants.

Mayo et al. [41] concluded that *A. palmeri* control generally decreased as application timing was delayed for acifluorfen, imazethapyr, and lactofen.

Herbicide/dose	Adjuvant[a]	Peanut injury[b,c] (%)	A. palmeri (%)	
		2012	2011	2012
Acifluorfen/0.28 kg/ha	None	3	7	22
	Agridex	7	30	30
	Induce	5	30	25
Acifluorfen/0.56 kg/ha	None	5	27	32
	Agridex	6	43	47
	Induce	6	43	47
Imazapic /0.035 kg/ha	None	0	33	70
	Agridex	0	22	67
	Induce	0	53	70
Imazapic /0.07 kg/ha	None	0	58	78
	Agridex	0	80	93
	Induce	0	58	93
Imazethapyr/0.035 kg/ha	None	0	27	10
	Agridex	0	27	27
	Induce	0	50	12
Imazethapyr/0.07 kg/ha	None	0	42	47
	Agridex	0	65	82
	Induce	0	27	67
Lactofen/0.11 kg/ha	None	4	8	15
	Agridex	4	18	22
	Induce	4	33	20
Lactofen/0.22 kg/ha	None	6	40	22
	Agridex	5	73	35
	Induce	6	38	37
2,4-DB/0.23 kg/ha	None	2	12	37
	Agridex	3	52	63
	Induce	3	65	60
2,4-DB/0.46 kg/ha	None	5	68	73
	Agridex	5	77	87
	Induce	5	78	83
LSD (0.05)		2	18	8

[a] *Adjuvant doses: Agridex, 1.0% v/v; Induce, 0.25% v/v.*
[b] *No injury was noted in 2011.*
[c] *Acifluorfen and lactofen leaf injury consisted of leaf burn, interveinal chlorosis, and marginal necrosis while 2,4-DB injury consisted of leaf curling and irregular leaf growth.*

Table 3.
Peanut injury and Amaranthus palmeri *control in the High Plains 1 month after herbicide application when using different adjuvants.*

Herbicide/dose	Adjuvant[a,b]	Peanut injury[c]		A. palmeri		C. melo	
		2012 (%)	2013 (%)	2012 (%)	2013 (%)	2012 (%)	2013 (%)
Imazapic/0.035 kg/ha	None	0	0	66	70	82	97
	Agridex	0	0	82	75	89	89
	Induce	0	0	72	86	91	82
	C-K II	0	0	66	82	85	93
	90–10	0	0	53	66	91	90
	ETA	0	0	88	80	94	97
Imazapic /0.07 kg/ha	None	0	0	75	64	99	95
	Agridex	0	0	73	71	97	91
	Induce	0	0	83	66	97	97
	C-K II	0	0	80	83	95	98
	90–10	0	0	88	79	99	99
	ETA	0	0	83	57	98	99
Lactofen/0.11 kg/ha	None	23	2	94	96	82	97
	Agridex	38	18	99	97	89	99
	Induce	27	4	91	92	91	95
	C-K II	33	18	99	100	85	94
	90–10	18	10	88	98	91	89
	ETA	33	19	93	96	94	99
Lactofen/0.22 kg/ha	None	22	6	97	98	99	100
	Agridex	38	20	97	89	97	100
	Induce	19	7	91	99	97	100
	C-K II	33	17	99	92	95	99
	90–10	23	10	99	95	99	99
	ETA	38	22	78	97	98	100
LSD (0.05)		11	5	22	12	14	9

[a]*Adjuvant dose: Agridex, 1.0% v/v; Induce, 0.25% v/v; Cide-Kick II, 1.0% v/v; 90–10, 1.0% v/v; and ETA, 1.0% v/v.*
[b]*C-K II, Cide-Kick II.*
[c]*Peanut injury ratings taken 4 days after herbicide application in 2012 and 7 days after herbicide application in 2013.*

Table 4.
Peanut injury, Amaranthus palmeri, *and* Cucumis melo *control in South Texas 1 month after imazapic and lactofen application when using different adjuvants.*

3.2 *Cucumis melo* L. control

In neither 2012 nor 2013 did the use of any adjuvant with either dose of imazapic or lactofen improve *C. melo* control over the use of no adjuvant (**Table 4**). In 2013, using Induce with the low dose of imazapic did reduce *C. melo* control compared to the use of no adjuvant or ETA. Imazapic at 0.04–0.07 kg/ha controlled greater than 90% *C. melo* in corn (*Zea mays* L.) regardless whether applied either preemergence, early POST or late POST [42], while Grichar [43] has seen similar results in peanut with imazapic POST applications.

The high humidities at application timing may have been a factor in the excellent control [44]. Wichert et al. [44] reported that relative humidity appeared to be a more important environmental factor than temperature on the activity of lactofen and other diphenylether herbicides on *Sida spinosa* L. Control of *Xanthium strumarium* L. and *Ambrosia artemisiifolia* L. with acifluorfen at 85% relative humidity was 10–30% greater than control with treatments applied at 50% relative humidity [45].

3.3 *Urochloa texana* control

The use of an adjuvant with either dose of clethodim did not improve *U. texana* control over clethodim alone at either evaluation timing or in either year (**Table 5**). Trends in 2013, when evaluated 2 weeks after herbicide application, did indicate that the addition of either Agridex or Cide-Kick II hastened the kill of *U. texana*. Jordan et al. [46] reported that the most consistent grass control with clethodim was obtained when applied with adjuvants containing a crop oil constituent or with the adjuvant Dash. They stated that although clethodim applied with a conventional nonionic adjuvant or silicone-based adjuvant controlled grasses in some instances, especially when applied at the higher dose of 0.14 kg/ha, control was inconsistent. They concluded that differences in efficacy among experiments could not be explained by differences in visible plant stress or extremes in temperature or relative humidity.

As seen with clethodim, the addition of an adjuvant to fluazifop-P-butyl did not improve *U. texana* control over the use of fluazifop alone at either dose (**Table 5**). At the time of herbicide application, relative humidity was at least 96% and soil moisture was excellent in both years (**Table 2**). These conditions can greatly influence herbicide activity [47, 48]. The effect of one climatic factor, such as humidity, will be greatest when other factors such as temperature or soil moisture are optimal [48].

3.4 Peanut injury

No injury was noted at the High Plains location in 2011 (data not shown). In some instances, imazapic can cause a yellowing of peanut plant for approximately 7–10 days after application; however, no injury was noted when rated 1 month after herbicide application (**Table 3**) in the High Plains or 4–7 days after application in South Texas (**Table 4**). Acifluorfen did cause a leaf burn, which was still noticeable 1 month after herbicide application (**Table 3**). The use of either Agridex or Induce with the lower dose of acifluorfen resulted in greater leaf burn than acifluorfen alone; however, this was not seen with the higher dose as there was no difference in leaf burn with/without the use of an adjuvant. The use of 2,4-DB did result in leaf curling and some irregular leaf growth, but no differences were noted with or without adjuvant with either dose (**Table 3**). Lactofen can also result in peanut leaf burn. Peanut injury ratings with lactofen were less in the High Plains (**Table 3**) than South Texas (**Table 4**), and this was due in part to the time interval between herbicide application ratings. In South Texas, ratings were taken 7 days or less after herbicide application, while in the High Plains, ratings were taken 30 days after application. Also, delaying the rating in 2013 by 3 days resulted in less injury than the 4 days evaluation (**Table 4**). At the High Plains location, no differences in leaf burn were noted with any adjuvant with either dose of lactofen (**Table 3**). At the South Texas location in 2012, leaf burn was greatest with Agridex when using the lower dose of lactofen (**Table 5**). At the higher dose of lactofen, Agridex, Cide-Kick II, and ETA resulted in greater injury than lactofen alone or lactofen plus Induce. In 2013, leaf burn with lactofen was greater when Agridex, Cide-Kick II, or ETA was used with either dose of lactofen. The addition of Induce or 90–10 resulted in 10% or less injury when added to either dose of lactofen (**Table 4**).

Herbicide/dose	Adjuvant[a,b]	*U. texana*			
		2 weeks after		4 weeks after	
		2012 (%)	2013 (%)	2012 (%)	2013 (%)
Clethodim/0.05 kg/ha	None	73	74	98	80
	Agridex	68	92	89	87
	Induce	75	77	97	88
	C-K II	60	82	84	83
	90–10	65	77	98	91
	ETA	77	81	98	70
Clethodim/0.1 kg/ha	None	84	90	98	95
	Agridex	91	90	99	94
	Induce	81	64	97	95
	C-K II	77	81	85	97
	90–10	76	77	93	94
	ETA	77	87	95	96
Fluazifop-P/0.11 kg/ha	None	65	81	92	96
	Agridex	62	85	65	93
	Induce	63	81	87	88
	C-K II	62	72	73	88
	90–10	63	95	78	97
	ETA	65	86	85	90
Fluazifop-P/0.22 kg/ha	None	67	88	92	95
	Agridex	58	62	91	93
	Induce	65	79	94	93
	C-K II	57	99	83	93
	90–10	62	96	95	97
	ETA	63	90	84	96
LSD (0.05)		15	24	19	15

[a]*C-K II, Cide-Kick II.*
[b]*Adjuvant dose: Agridex, 1.0% v/v; Induce, 0.25% v/v; C-K II, 1.0% v/v; 90–10, 1.0% v/v; and ETA, 1.0% v/v.*

Table 5.
U. texana *control with clethodim and fluazifop-P when using different adjuvants.*

4. Conclusion

The use of an adjuvant in South Texas did not always improve weed efficacy, while in the High Plains of Texas, the use of an adjuvant did improve weed efficacy in most instances. The herbicide-adjuvant-plant interaction is a complex system. An adjuvant can impose its impact at several stages of the herbicide application including tank mixing, deposition and retention on the plants, absorption by the plants, and translocation from the applied area to the site of action [15, 18, 20, 23]. Understanding the different roles of adjuvants in enhancing herbicide efficacy is essential for the optimum use of adjuvants in herbicide application. Reducing the herbicide rate proved to be effective in South Texas but not so in the High Plains due to several factors including a higher relative humidity, the time of herbicide application in the early morning hours, and excellent moisture conditions at time of herbicide application in South Texas. Postemergence herbicide efficacy may be affected by environmental factors including light duration and intensity, air temperature, relative humidity, and dew or precipitation [47–51]. These environmental conditions may influence processes such as herbicide absorption, translocation, or plant metabolism, which influence herbicide efficacy [52]. Air temperature in South Texas varied from 25 to 31°C, while air temperature varied from 29 to 32°C in the High Plains region (**Table 2**). As temperature increased, glyphosate efficacy on *Avena fatua* L., *Urochloa panicoides* Beauv. [53], and *Echinochloa colona* (L.) Link [54] increased. Temperature also influenced ^{14}C-glyphosate absorption by cultured velvetleaf cells [55]. Nearly twice as much glyphosate was absorbed at 28°C than at 4 or 16°C. Similar temperature effects were observed with *Cynodon dactylon* (L.) Pers. [56] and *Sorghum halepense* L. [57]. Herbicide activity or absorption increased with increasing relative humidity for *Elytrigia repens* (L.) Nevski [58], *C. dactylon* [56], *U. panicoides* [53], *E. colona* [54], and *S. halepense* [57]. Generally, high relative humidity and high temperatures, as well as low light intensity before treatment, increased plant susceptibility to POST herbicides [49]. Plant stress may also reduce systemic herbicide activity and account for relatively poor performance. Buhler and Burnside [58] noted that glyphosate was less effective on drought-stressed annual grass species than actively growing plants. Contact herbicides such as lactofen are not as dependent on translocation for activity, and their activity is not as adversely affected by drought-stressed plants. The above-mentioned factors all contributed to the lack of difference of the postemergence herbicides alone or with an adjuvant as well as the effectiveness of the 1/2X herbicide doses specifically in the South Texas studies. Many field applications of herbicides in South Texas start early in the morning to avoid windy conditions that may develop late in the day when coastal sea breezes may start up. Under the early morning conditions, with high humidity, dew can be found on weeds as well as on the crop at the time of application. Dew, defined as the presence of free water on plant foliage [59], could affect the foliar uptake and therefore efficacy of foliar-applied herbicides, mainly those of high water solubility. The presence of dew at application is believed to increase or decrease foliar herbicide efficacy [60]. Herbicide runoff and herbicide dilution could explain the negative effect of dew [61]. By contrast, dew can increase the total area of herbicide interception and reduce the impact of large drops on foliage surfaces, avoiding their loss from the leaves [62]. At the same time, the presence of dew results in hydration of the cuticle and may play an important role favoring foliar uptake [59]. The effect of dew on herbicide activity is not thoroughly understood, due to limited research where dew has not been quantified [63]. Another factor which may explain the lack of a response to a surfactant in South Texas may be the effects of higher spray volumes used in South Texas (190 L/ha) compared to the High Plains (94 L/ha). In order to obtain acceptable control with lactofen, a contact herbicide, a large portion of the leaf, must receive a spray solution [64] and with higher spray volumes in South Texas more of the leaflet received spray coverage. The drift guard (DG) nozzles used in South Texas contained larger droplets [65]. A larger droplet size causes localized injury to the weed leaf resulting in better control with contact herbicides such as acifluorfen and lactofen. Several researchers have investigated the effects of carrier volumes on the efficacy of herbicides [66–69] and results have been variable. At a constant spray droplet size, glyphosate [66, 67] and paraquat efficacy increased as application volume decreased. However, clopyralid activity decreased as application volume decreased [68]. Results from these trials suggest that not all adjuvants perform the same for individual herbicides. It is critical that a quality adjuvant be used when the label suggests that one is needed for maximum herbicidal activity. Since adjuvants may also increase herbicidal toxicity to crops, it is also critical to omit the surfactant if the label suggests to do so for individual herbicides.

Acknowledgements

The Texas Peanut Producers Board and the National Peanut Board provided financial support for this research. Dwayne Drozd provided technical assistance.

Conflict of interest

The authors have declared that no competing interests exist.

Author details

William James Grichar[1*], Peter A. Dotray[2] and Mark A. Matocha[3]

1 Texas A&M AgriLife Research, Corpus Christi, TX, USA

2 Texas A&M AgriLife Research and Texas A&M AgriLife Extension Service, Lubbock, TX, USA

3 Texas A&M AgriLife Extension Service, College Station, TX, USA

*Address all correspondence to: w-grichar@tamu.edu

References

[1] Tu M, Hurd, C, Randall JM. Adjuvants. Weed Control Methods Handbook: Tools & Techniques for Use in Natural Areas. The Nature Conservancy. version: April 2001. pp. 8.1-8.25. Available from: http://tncweeds.ucdavis.edu [Accessed: October 31, 2018]

[2] Klingaman TE, King CA, Oliver LR. Effect of application rate, weed species, and weed stage of growth on imazethapyr activity. Weed Science. 1992;**40**:227-232

[3] York AC, Jordan DL, Wilcut JW. Effects of $(NH_4)_2SO_4$ and BCH 81508 S on efficacy of sethoxydim. Weed Technology. 1990;**4**:76-80

[4] Holshouser DL, Coble HD. Compatibility of sethoxydim with five postemergence broadleaf herbicides. Weed Technology. 1990;**4**:128-133

[5] Kent LM, Willis GD, Shaw DR. Influence of ammonium sulfate, imazapyr, temperature, and relative humidity on the absorption and translocation of imazethapyr. Weed Science. 1991;**39**:412-416

[6] Hatzios KK, Penner D. Interactions of herbicides with other agrichemicals in higher plants. Review Weed Science. 1985;**1**:1-63

[7] Kammier KJ, Walters SA, Young BG. Effects of adjuvants, halosulfuron, and grass herbicides on *Cucurbita* spp. injury and grass control. Weed Technology. 2010;**24**:147-152

[8] Bridges DC. Adjuvant and pH effects on sethoxydim and clethodim activity on rhizome johnsongrass (*Sorghum halepense*). Weed Technology. 1989;**3**:615-620

[9] Wanamarta G, Penner D. Foliar penetration of herbicides. Review Weed Science. 1989;**4**:215-231

[10] Knezevic SZ, Datta A, Scott J, Charvat LD. Adjuvants influenced saflufenacil efficacy on fall-emerging weeds. Weed Technology. 2009;**23**:340-345

[11] McWhorter CG. The use of adjuvants. In: Hodson RH, editor. Adjuvants for Herbicides. Champaign, IL: Weed Science Society of America; 1992. pp. 10-25

[12] Van Valkenburg JW. Terminology, classification, and chemistry. In: Hodson RH, editor. Adjuvants for Herbicides. Champaign, IL: Weed Science Society of America; 1992. pp. 1-3

[13] Hull HM, Davis DG, Stolzenberg GE. Action of adjuvants on plant Surfaces. In: Hodson RH, editor. Adjuvants for Herbicides. Champaign, IL: Weed Science Society of America; 1992. pp. 26-67

[14] DiTomaso JM. Barriers to foliar penetration and uptake of herbicides. Proceedings of the California Weed Science Society. 1999;**51**:150-155

[15] Singh M, Tan S, Sharma SD. Adjuvants enhance weed control efficacy of foliar-applied diuron. Weed Technology. 2002;**16**:74-782

[16] Kocher H, Kocher J. Influence of wetting agents on the foliar uptake and herbicidal activity of glufosinate. Pesticide Science. 1993;**37**:155-158

[17] Singh M, Orsenigo JR, Shah DO. Surface tension and contact angle of herbicide solutions affected by surfactants. Journal American Oil Chemist Society. 1984;**61**:596-599

[18] Foy CL, Smith LW. Surface tension lowering, wettability of paraffin and corn leaf surfaces, and herbicidal enhancement of dalapon by seven surfactants. Weeds. 1965;**13**:15-19

[19] Kirkwood RC. Recent developments in our understanding of the plant cuticle as a barrier to the foliar uptake of pesticides. Pesticide Science. 1999;**55**:69-77

[20] Anonymous. Influence of adjuvants on absorption of foliar-applied herbicides. Plant & Soil Sciences eLibrary. Lincoln: University of Nebraska. Available from: https://passel.unl.edu/pages/informationmodule.php?idinformationmodule+1056648673&topicorder=3&maxto=8&minto=1 [Accessed: September 09, 2014]

[21] Singh M, Mack RE. Effect of organosilicone-based adjuvants on herbicide efficacy. Pesticide Science. 1993;**38**:219-225

[22] Stevens PJG. Organosilicone surfactants as adjuvants for agrochemicals. Pesticide Science. 1993;**38**:103-122

[23] Stevens PJG, Kimberley MO, Murphy DS, Policello GA. Adhesion of spray droplets to foliage: The role of dynamic surface tension and advantages of organosilicone surfactants. Pesticide Science. 1993;**38**:237-245

[24] Simpson CE, Baring MR, Schubert AM, Melouk HA, Lopez Y, Kirby JS. Registration of 'Olin' peanut. Crop Science. 2003;**43**:1880-1881

[25] Baring MR, Simpson CE, Burow MD, Black MC, Cason JM, Ayers J, et al. Registration of 'Tamrun OL07' peanut. Crop Science. 2006;**46**:2721-2722

[26] Beasley J, Baldwin J. Peanut cultivar options and descriptions, 2009. Available from: http://www.uga/commodities/fieldcrops/peanuts/production/cultivardescription.html [Accessed: September 09, 2018]

[27] Frans RR, Talbert R, Marx D, Crowley H. Experimental design and techniques for measuring and analyzing plant responses to weed control practices. In: Camper ND, editor. Research Methods in Weed Science. 3rd ed. Champaign, IL: Southern Weed Science Society; 1986. pp. 29-46

[28] SAS Institute Incorporated. SAS/STAT User's Guide: Statistics, Version 9.1. 2007. Cary, IN:SAS Institute; 2007

[29] NWS, Lubbock, TX. A Record Breaking Year! National Weather Service Weather Forecast Office. 2011. Available from: http://www.srh.noaa.gov/lub/?n=events-2011-2011-record-countdown [Accessed: October 10, 2018]

[30] DeFelice MS, Brown WB, Aldrich RJ, Sims BD, Judy DT, Guethle DR. Weed control in soybeans (*Glycine max*) with reduced rates of postemergence herbicides. Weed Science. 1989;**37**:365-374

[31] Devlin LD, Long JH, Maddox LD. Using reduced rates of postemergence herbicides in soybeans (*Glycine max*). Weed Technology. 1991;**5**:834-840

[32] Hamill AS, Zhang JH. Rate and time of bentazon/atrazine application for broadleaf weed control in corn (*Zea mays*). Weed Technology. 1997;**11**:549-555

[33] Gorski SF, Wertz MK. Tomato (*Lycopersicon esculentum*) and Eastern black nightshade (*Solanum ptycanthum*) tolerance to acifluorfen. Weed Technology. 1987;**1**:278-281

[34] Lee SD, Oliver LR. Efficacy of aciflurofen on broadleaf weeds: Time and methods for application. Weed Science. 1982;**30**:520-526

[35] Buhler DD, Doll JD, Proost RT, Visocky MR. Integrating mechanical weeding with reduced herbicide use in conservation tillage corn production systems. Agronomy Journal. 1995;**87**:507-512

[36] Stougaard RN, Maxwell BD, Harris JD. Influence of application timing on the efficacy of reduced rate postemergence herbicides for wild oat (*Avena fatua*) control in spring barley (*Hordeum vulgare*). Weed Technology. 1997;**11**:283-289

[37] Levene BC, Owen MDK. Effect of moisture stress and leaf age on bentazon absorption in common cocklebur (*Xanthium strumarium*) and velvetleaf (*Abutilon theophrasti*). Weed Science. 1995;**43**:7-12

[38] Coudercher M, Retzlaff G. Daily changes in the relative water content of velvetleaf (*Abutilon theophrasti* Medic.) may explain its rhythmic sensitivity to bentazon. Journal of Plant Physiology. 1995;**145**:501-506

[39] Bellinder RR, Arsenovic M, Shab DA, Rauch BJ. Effect of weed growth stage and adjuvant on the efficacy of fomesafen and bentazon. Weed Science. 2003;**51**:1016-1021

[40] Grichar WJ, Dotray PA. Controlling weeds found in peanut with lactofen. Crop Management. 2011. DOI: 10.1094/CM-2011-0912-01-RS

[41] Mayo CM, Horak MJ, Peterson DE, Boyer JE. Differential control of four Amaranthus species by six postemergence herbicides. Weed Technology. 1995;**9**:141-147

[42] Thompson AM, Rosales-Robles E, Chandler JM, Nester PR, Tingle CH. Crop tolerance and weed management systems in imidazolinone-tolerant corn (*Zea mays* L.). Weed Technology. 2005;**19**:1037-1044

[43] Grichar WJ. Herbicide systems for control of horse purslane (*Trianthema portulacastrum* L.), smellmelon (*Cucumis melo* L.), and Palmer amaranth (*Amaranthus palmeri* S. Wats) in peanut. Peanut Science. 2008;**35**:38-42

[44] Wichert RA, Bozsa R, Talbert RE, Oliver LR. Temperature and relative humidity effect on diphenyl herbicides. Weed Technology. 1992;**6**:19-24

[45] Ritter RL, Coble HD. Penetration, translocation, and metabolism of aciflurofen in soybean (*Glycine max*), common ragweed (*Ambrosia artemisiifolia*), and common cocklebur (*Xanthium pensylvanicum*). Weed Science. 1981;**29**:474-478

[46] Jordan DL, Vidrine PR, Griffin JL, Reynolds DB. Influence of adjuvants on efficacy of clethodim. Weed Technology. 1996;**10**:738-743

[47] Gerber HR, Nyffeler A, Green DH. The influence of rainfall, temperature, humidity, and light on soil- and foliar-applied herbicides. Aspects of Applied Biology. 1983;**4**:1-14

[48] Kudsk P, Olesen T, Thonke KE. The influence of temperature, humidity and simulated rainfall on the performance of thiameturon-methyl. Weed Research. 1990;**30**:330-335

[49] Hammerton JL. Environmental factors and susceptibility to herbicides. Weeds. 1967;**15**:330-335

[50] Price CE. The effect of environment on foliage uptake and translocation of Herbicides. Aspects of Applied Biology. 1983;**4**:157-169

[51] Muzik TJ. Influence of environmental factors on toxicity to plants. In: Audus LJ, editor. Herbicides: Physiology, Biochemistry, Ecology. Vol. I. London: Academic Press; 1976. pp. 58-64

[52] Waltz AL, Martin AR, Roeth FW, Lindquist JL. Glyphosate efficacy on velvetleaf varies with application time of day. Weed Technology. 2004;**18**:931-939

[53] Adkins SW, Tanpipat S, Swarbrick JT, Boersma M. Influence of

environmental factors on glyphosate when applied to *Avena fatua* or *Urochloa panicoides*. Weed Research. 1998;**38**:129-138

[54] Tanpipat S, Adkins SW, Swarbrick JT, Boersma M. Influence of selected environmental factors on glyphosate efficacy when applied to awnless barnyardgrass (*Echinochloa colona (L.)* Link). Australian Journal of Agriculture. 1997;**48**:695-702

[55] Reyneberg T, Balke NE, Lund-Hoie K. Effects of adjuvants and temperature on glyphosate absorption by cultured cells of velvetleaf (*Abutilon theophrasti* Medic.). Weed Research. 1992;**32**:419-428

[56] Jordan TN. Effects of temperature and relative humidity on the toxicity of glyphosate to bermudagrass (*Cynodon dactylon*). Weed Science. 1977;**25**:448-451

[57] McWhorter CG, Jordan TN, Wills GD. Translocation of ^{14}C-glyphosate in soybeans (*Glycine max*) and johnsongrass (*Sorghum halepense*). Weed Science. 1980;**28**:113-118

[58] Buhler DD, Burnside OC. Effect of spray components on glyphosate toxicity to annual grasses. Weed Science. 1983;**31**:124-130

[59] Slatyer R. Environmental aspects of plant weather relationship. In: Plant-Weather Relationship. 2nd ed. New York: Academic Press; 1968. pp. 40-46, 230-236

[60] Caseley JC. Variation in foliar pesticide performance attributable to humidity, dew, and rain effects. Aspects of Applied Biology (CAB abstract). 1989;**21**:215-225

[61] Muro J. Influencia del Rocio y del volume de caldo de aplicación en la eficacia de tres herbicidas de contacto. Córdoba: Actas Reunión Society, Espanola de Malerbologia; 1991. pp. 271-274

[62] Johnstone D. Spreading and retention of pesticide sprays. In: van Valkenburg NW, editor. Pesticide Formulations. New York: Marcel Dekker; 1973. pp. 343-386

[63] Kogan M, Bayer ED. Herbicide uptake as influenced by plant water status. Pesticide Biochemistry and Physiology. 1996;**56**:174-182

[64] McKinley KS, Ashford R, Ford RJ. Effects of drop size, spray volume, and dosage on paraquat toxicity. Weed Science. 1974;**22**:31-31

[65] Anonymous. Teejet Technologies. 2018. Available from: http://www.teejet.com/english/home/products/spray-products/broadcast-spray-nozzles-turbo-tee [Accessed: October 21, 2018]

[66] Ambach RM, Ashford R. Effects of variations in drop makeup on the phytotoxicity of glyphosate. Weed Science. 1982;**30**:221-224

[67] Buhler DD, Burnside OC. Effects of application variables on glyphosate phytotoxicity. Weed Technology. 1987;**1**:14-17

[68] Cranmer JR, Linscott DL. Droplet makeup and the effect on phytotoxicity of glyphosate in velvetleaf (*Abutilon theophrasti*). Weed Science. 1990;**38**:406-410

[69] Whisenant SG, Bouse LF, Crane RA, Bovey RW. Droplet size and spray volume effects on honey mesquite mortality with clopyralid. Journal of Range Management. 1993;**46**:257-261

Chapter 8

Breeding Elite Cowpea [*Vigna unguiculata* (L.) Walp] Varieties for Improved Food Security and Income in Africa: Opportunities and Challenges

Ana Maria Figueira Gomes, Nascimento Nhantumbo, Manuela Ferreira-Pinto, Rafael Massinga, José C. Ramalho and Ana Ribeiro-Barros

Abstract

Cowpea, *Vigna unguiculata* (L.) Walp, is among the most important grain legumes in Africa. Its nutritional value and biological nitrogen fixation (BNF) potential coupled with a high plasticity to environmental conditions places this legume in a unique position in Sub-Saharan Africa (SSA) in the context of food and nutritional security. However, cowpea yield and BNF contribution to agricultural systems in this sub-continent is far behind the average global values. The inability to run effective breeding programs to timely generate and deliver high yielding, nutritious and climate smart cowpea varieties, coupled with poor crop husbandry practices has been in the forefront of the current situation. In this chapter, the main constrains and opportunities to establish and run successful and effective cowpea production and breading programs in SSA are discussed. The discussion is built around the argument that SSA can benefit from its rich collection of landraces, as well as from high-throughput methodologies to assist the screening and the development of adapted, high yielding and nutritious varieties.

Keywords: cowpea, breeding, food security, Africa

1. Introduction

Cowpea, *Vigna unguiculata,* is a legume crop widely regarded as the "poor men's meat", due to the high protein contents in leaves, pods and grains [1]. Besides that, cowpea presents high plasticity which allows it to thrive under a wide range of environmental conditions [2]. These characteristics, together with its biological nitrogen fixation (BNF) capacity in symbiosis with rhizobia bacteria, make cowpea an important crop to rural households from Sub-Saharan Africa (SSA), whose diet is mainly based on carbohydrate rich crops and agricultural systems are largely deficient. Despite the fact that SSA is among the main cowpea producers and primary consumers, its yield and BNF return is the lowest when compared with the rest of the world [3, 4]. In consequence, the sub-continent's production is far from satisfying the internal demand.

With the exponential growth of the world's population, which is anticipated to be *ca*. 10 billion by 2050 [5], 60% of which in Africa, the demand for food in the continent is anticipated to grow by as much as 400% [6]. Taking into consideration the current scenarios of climate changes and the predictions for the middle of this century, *i.e.*, a high probability for the occurrence of temperature and CO_2 increases, coupled with altered rainfall patterns and soil salinity [7–9], the impact of population growth on food and nutritional security will be further exacerbated. Given this reality, the design and promotion of climate-smart food systems will be mandatory to achieve most of the United Nations Sustainable Development Goals [10]. Thus, accelerating the development and implementation of a nutrition-sensitive agricultural research and development agenda, particularly in making the breeding programs in SSA more responsive to its nutritional and agro-ecological context will be more relevant than ever. In this chapter the main cowpea production constraints in SSA are discussed, bringing forward the major challenges and opportunities to breed elite cowpea varieties towards self-sufficiency and competitiveness in the global arena.

2. Cowpea in sub-Saharan Africa

2.1 Importance and potential contribution to better diets and food security

In most developing countries from SSA cowpea is the most accessible nutritional source [11]. The leaves for instance, are more nutrient-dense than many other leaf vegetables [12, 13]. Cowpea is also a source of minerals and vitamins [14]. High lysine content of grain proteins plays a key role in balancing cereals and cassava-based diets, typical of most African countries [15]. Additionally, low fat and high carbohydrate contents make cowpea a balanced food source [16]. An analysis of 1541 cowpea germplasm lines [17] revealed that on average cowpea has 25% protein and *ca.* 38 mg Zn/kg, 53 mg Fe/kg, 1.9 g Mg/kg, 0.825 g Ca/kg, 5 g P/kg, and 15 g K/kg. Cowpea plays also an important role in soil nutrient cycling [18] as a result of its capacity to establish N_2-fixing root-nodule symbiosis with rhizobia bacteria. In modern agriculture systems, cowpea can contribute with 70–350 kg nitrogen per ha through biological nitrogen fixation (BNF) [19]. Thus, it is an important resource management technology in cereal-based systems leading to *ca.* three-fold yield increases of unfertilized maize [20–22].

2.2 Biotic stress: pests, diseases and weeds

One of the reasons associated with the low cowpea yields in SSA is the impact of several pests (**Table 1**). Aphids (*Aphis craccivora* Koh) are among the main pests affecting cowpea production, particularly at the seedling stage [23]. However, the impact can be minimized through the use of tolerant cultivars coupled with proper agronomic management procedures [33]. Another major threat to cowpea is posed by post flowering and podding pests, such as the flower thrips (*Megalurothrips sjostedti* Trybom), the legume pod borer (*Maruca vitrata* Fab.) and pod sucking bugs from the Hemiptera order, of which *Clavigralla tomentosicollis* Stal is the most important in tropical Africa [34]. In severely infested fields, post flowering pests can lead up to 70–80% yield loss [35]. Several measures have been used to minimize the impact of these pests, including pesticides, genetically modified (GM) varieties, as well as integrated pest management (IPM) practices [36].

Species (order: family)	Plant part attacked	Importance	Reference
Aphis craccivora Koch (Homoptera: Aphididae)	Leaves, flowers and pods	Key	[23–25]
Empoasca dolichi Paoli (Homoptera: Cicadelidae)	Leaves	Sporadic	[26]
Ophiomyia phaseoli (Tryon) (Diptera: Agromyzidae)	Stem	Sporadic	[27]
Amsacta moorei (Butler) (Lepidoptera: Arctiidae)	Leaves	Sporadic	[28]
Megalurothrips sjostedti (Trybom) (Thysanoptera: Thripidae)	Floral structures	Key	[24, 29]
Maruca vitrata (Fab.) (Lepidoptera: Pyralidae)	Stem, flowers, pods	Key	[24, 29]
Clavigralla tomentosicollis Stal (Hemiptera: Coreidae)	Pods	Key	[24, 29]
Riptortus dentipes (Fab.) (Hemiptera: Alydidae)	Pods	Sporadic	[28]
Nezara viridula Linnaeus (Hemiptera: Pentatomidae)	Pods	Sporadic	[28]
Callosobruchus spp. (Coleoptera: Bruchidae)	Seeds (storage)	Key	[30–32]

Table 1.
Major field and storage pests of cowpea: Attacked plant parts and importance.

The first GM pod borer resistant (PBR) cowpea was introduced in Nigeria in 2011 [37–39], and then expanded to Burkina Faso [39], Ghana [40], and Malawi [39]. However, results are still preliminary and most countries with on-going trials are yet to release GM-PBR cowpea, pending the evidence on GM cowpea performance, as well as the legal issues, such as competition with non-GM landraces, and assess of smallholder farmers to transgenic seeds [39]. Therefore, the GM option needs to be part of a feasible integrated IPM package that can easily meet local farmers' needs and capacities while offering an easily accessible solution.

Callosobruchus maculatus (Fab.), a cosmopolitan storage pest, is one of the most important off-the field pests affecting African cowpea producers mainly due to poor post-harvest storage conditions [30]. The attack normally leads to weight loss, decreased retail and nutritional value and reduced seed germination rate [27, 41]. So far, chemical control coupled with the use of resistant varieties have offered the best response to resource endowed smallholder cowpea producers across SSA, which also use grain hardness as a key selection trait to reduce storage losses [42–44]. More recently, hermetic grain storage technologies have been promoted [44–46]. However, these technologies are yet to reach most resource poor farmers.

Besides pests, cowpea is also susceptible to several fungal, bacterial and viral diseases. Bacterial blight caused by *Xanthomonas axonopodis* (Smith) is the most damaging bacterial disease [47]. This seed-borne disease can lead to almost 60% seedling mortality and can survive on crop residues [27]. Therefore, the use of healthy seeds and resistant varieties is the best option to control the disease [48] . On the other hand, cowpea anthracnose caused by *Colletotrichum lindemuthianum* (Sacc. & Magn.) is the leading fungal disease, mainly during cool and wet weather [41]. Yield losses of 30–50% have been reported in highly susceptible lines grown in monocrops where the disease attack is most severe and the agent spreads easily [49].

Viruses have been even more problematic than fungal and bacterial diseases, thus needing particular attention [41, 50]. In total, eight major viral diseases were reported to affect cowpea in SSA. These can be divided in four groups based on the main propagation agent. Three are beetle-transmitted, namely, the cowpea yellow mosaic virus (CYMV), cowpea mottle virus (CMV) and southern bean mosaic virus (SBMV); two aphid-borne viruses namely, the cowpea aphid-borne mosaic potyvirus (CABMV) and cucumber mosaic cucumovirus (CMV); and two whitefly-transmitted viruses namely, cowpea golden mosaic virus (CGMV) and cowpea mild-mottle carlavirus (CPMMV). The eighth disease, whose agent is unknown to date, is the sunn-hemp mosaic virus (SHMV) [51], a tobamovirus that attacks several legume species [52]. Of the eight viruses, CABMV is the most problematic. In Nigeria, Oderara and Kumar [53] and Shoyinka and collaborators [54] analyzed 315 and 649 cowpea lines, respectively, and found that CABMV had high incidence across all sampled agroecological regions with up to 64% yield losses. Recently, Mukoye and collaborators [55] reported yield losses ranging from 10–100% in Western Kenya. The use of clean seeds and resistant varieties are the most cost-effective practices to control viruses [55], but recent research has shown promising results with IPM and the use of plant extracts in controlling the transmission agents, *i.e.*, pests [56]. The use of allelopathic effects, a technology that has gained prominent use to manage field pests in Asia and Latin America [57–59] is also another alternative to be explored in Africa. Trap cropping [56], a well-known strategy to manage insect pest through diversification of the plant strata to stimulate the population of natural enemies is also a practice to be massified.

Weeds also present a serious problem to cowpea mainly during crop establishment when more attention towards weed control is required [60]. At this stage severe competition for light, nutrient and space are responsible for considerable reduction in crop yield [61]. The parasitic weeds, *Striga gesnerioides* (Willd.) Vatke ex Engl. and *Alectra vogelii* Benth. are the major limitations to cowpea production in Africa, particularly in the dry savannas of West and Central Africa, *i.e.,* Sudan, Sahel and Guinea and portions of eastern and southern Africa [11, 62]. In total, yield losses between 73 and 100% by *S. gesnerioides* infestations have been reported in Africa [63]. Breeding efforts to transfer the *Bt*-gene to cowpea as a way to reduce the incidence of striga are ongoing with an *ex ante* economic impact assessment in West and Central Africa estimated in $1.2, $3.1 and $8.4 billion dollars in Benin, Niger and Nigeria respectively [64]. However, no Bt-cowpea has been available commercially in the region so far.

2.3 Abiotic stress: Water, nutrients and heat

Abiotic factors, such as, high temperature, drought and soil fertility are of upmost importance to plant development. Environmental stressors can lead to considerable cowpea yield losses in most SSA rain-fed agricultural systems. In the African dry savannas, characterized by hot days with high temperatures (above 35°C) spread across a short growing season, flower abortion and infertility due to poor pollen development is a common cause of yield reduction [11]. Singh and collaborators [65], observed that cowpea plants exposed to temperatures of 30–38°C, from 8 days after emergency to maturity, had a limited vegetative growth and reproductive potential. However, heat tolerant genotypes were able to retain flower production with a greater pod set [66].

Cowpea is frequently considered as a drought tolerant crop, linked also to the nitrogen fixing capacity of symbiotic rhizobia bacteria. However, in SSA where most systems are rain-fed, drought caused mainly by deficit of rainfall for long time periods has been a major threat to cowpea production [67, 68]. Ibrahim and collaborators [69] reported significant decreases in biomass production and water use efficiency (WUE) in six Ghanaian varieties subjected to water stress. Additionally, Fatokun and collaborators [1] observed that drought delayed the flowering process in 12 days and consequently the grain yield in *ca.* 70%. This might be explained by the decrease in leaf area and the concomitant photosynthetic rate and stomatal conductance [67].

One solution is the use of water efficient varieties coupled with better crop husbandry practices. The on-going efforts to screen and breed for drought tolerance and water efficient varieties, attaining more grain per drop, are essential in the African context where the crop is mostly cultivated under rainfed conditions and frequently exposed to intermittent droughts [68]. Thus, the use of well adapted early maturity cultivars seems to be one of the best solutions for smallholder cowpea producers to escape the effects of late season droughts [11].

Soil nutrient imbalances, particularly phosphorous (P) and nitrogen (N) have deserved less attention in cowpea research, despite the BNF potential to improve nutrient cycling and yields in African low external input agricultural systems [18]. According to Jemo and collaborators [70], BNF was significantly reduced in soils with low P levels and limited water supply. The same authors observed that as the level of P increased there was a significant reduction of water-deficit associated damages on BNF potential. Research has also demonstrated that supplying non-nodulated cowpea varieties with small nitrogen doses, promoted branching and increased crop yield [1].

3. Cowpea breeding programs in SSA: History, challenges and opportunities

Worldwide, cowpea breeding programs have targeted qualitative and quantitative traits to enhance the crop productive performance. The primordial breeding programs (1960–1980's) in SSA focused on high grain yield and seed quality, maturity time (extra-early, early and late), light sensitivity (photo-insensitive), growth habit (erect), intercrop fitting, lodging, and pest and disease resistance [1]. This was done mainly through a conventional breeding pipeline that included mainly germplasm collection, evaluation, maintenance and screening for desired traits mostly in Nigeria, Senegal, Uganda and Tanzania. Nowadays, breeding for drought tolerance [71, 72] and pest and disease resistance [73–76] have deserved major attention where the use of genomic tools is slowly gaining space. The International Institute of Tropical Agriculture (IITA) in Nigeria, and its international partners have played a key role in cowpea research and breeding initiatives. The Semi-Arid Food Grains Research and Development (SAFGRAD) project in the 1980's and more recently the Tropical Legumes project (2007–2018) and the CGIAR Cowpea Genomics Initiative (2005) marked a new step in cowpea breeding in SSA. Despite this, the number of varieties released in SSA is still small and there are more promising breeding lines than officially released varieties. In total, 80 IITA supported cowpea varieties were released, 24 of which during the past decade in 13 out of 54 African countries.

Despite the referred efforts, there are several constraints to cowpea breeding programs in SSA, which can be attributed to several factors, namely:

I. Poor investments in agricultural Research and Develop- ment (R&D) at national level and departmentalization of breeding programs: IITA and National Agricultural Research Systems (NARS) have been in the forefront of much breeding efforts in SSA, but the involvement of the regional agricultural universities (AUs) is not consolidate. In fact, only in Nigeria, Senegal, Uganda, Ghana, Tanzania and Kenya university-based research has been reported [1]. In addition to that, R&D in private sector is practically inexistent in SSA. Therefore, the region would benefit from a collaborative approach between international and regional R&D institutions (including AUs) and NARS, promoting the internationalization of the local R&D systems regarding scientific and technical work and publications, and engaging competitive funding raising.

II. Nutrition-sensitive trait selection for improved dietary quality: interest in cowpea's nutrition quality in Africa is an old issue [76], but it has been overlooked over the years. However, with the continent's nutrition agenda becoming increasingly important, a targeted breeding agenda on the nutritional quality of the crop is needed [12, 77]. Contrarily to Africa, the production of varieties with high dietary quality has deserved much attention in Europe and Asia [78, 79]. Currently, screening segregating populations for traits such as Fe, Zn, Cu and Mo content is in progress [80]. Such efforts are essential to improve the crop's contribution to local diets, as well as for the establishment of nutrition sensitive food systems. Special attention should be also given to fresh leaves and pods rather than solely focusing on dry pods and grain as it has happened so far. Increasing protein and mineral content, the latter also through biofortification, needs to be on top of the agenda.

III. Breeding approach: most breeding programs in Africa rely on open environment conventional breeding technics centered mostly on single trait selection methods. However, in the developing world, molecular characterization of germplasm, based in modern genomics and molecular marker-assisted selection [1] and genetic engineering [80, 81] coupled with digital imaging in high-throughput phenotyping [82], historical data [83, 84] and model-assisted selection [84–87] have revolutionized crop breeding programs. Such approach facilitated molecular, morpho-agronomic, physiological and biochemical characterization of cowpea germplasm to identify the best performing genotypes [88]. This integrative screening and selection approach represents a clear shift from single-trait to multiple-trait selection [85], something that is scantly done in African screening programs. By doing multiple-trait selection, the effectiveness and efficiency of breeding programs have been significantly improved in Europe, America and Australia, where significant investments in research infrastructure and human resource training has been made [89]. Model assisted breeding has proved to be fundamental in helping underpin prediction of likely phenotypical consequences of trait and genetic variations in targeted environments [86]. Furthermore, the agricultural production simulator (APSIM) has been successfully used in phenotyping and evaluating Genotype × Environment × Management (G × E × M) effects on drought adaptation. The growing interest in genotype-to-phenotype (G2P) models which predict phenotypic traits as a function of genotypic and environmental inputs is currently helping to enhance phenotype screening [89]. Additionally, the use of speed breeding chambers (SBC) [90], is also a recent and important advance in breeding programs. Such facilities allowed breeders to achieve up to six generations per year from spring wheat, durum wheat, barley, pea, chickpea and groundnuts, instead of one to three generations per year usually possible under field conditions and glasshouse, respectively [91].

IV. Improve cross-country coordination mechanisms and systematization of existing information: over the last decades several projects involving cowpea landraces screening and the assessment of their genetic diversity have been conducted in Africa [77]. However, the knowledge generated from this research is scattered all over the region and needs to be systematized and made available to aid current and future breading programs. For that to happen, cross-country coordination mechanisms and collaborative research opportunities need to be improved.

4. Conclusion

With an increasing world population, there is an urgent need to re-structure the R&D agenda in SSA towards the development of elite crop varieties that are more likely to successfully cope with future climate conditions. Cowpea, despite its high plasticity to survive in harsh environments, will not be an exception. The crop's importance in SSA as a food crop, animal feed and nutrient cycling agent makes it a candidate crop for future improvement and to operationalize the continents' nutrition agenda. For that, coordinate R&D efforts should be made at the regional level, in order to: (i) address the best production and breeding practices, through a wide screening of landraces towards the identification of the best performing genotypes (yield and nutritional quality) under limiting environmental conditions; (ii) identify multiple breeding traits and molecular tools for marker-assisted selection; and (iii) develop fast and reliable methods for variety certification, linked to important investment in R&D facilities and advanced training of human resources.

Acknowledgements

The authors acknowledge the support of Fundação para a Ciência e a Tecnologia (FCT), Portugal, through grant SFRH/BD/113952/2015 and research units UID/AGR/04129/2013 (LEAF) and UID/GEO/04035/2013 (GeoBioTec), and the Dutch Organization for Internationalization in Education (NUFFIC), through the project NICHE-MOZ-151.

Author details

Ana Maria Figueira Gomes[1,2], Nascimento Nhantumbo[2], Manuela Ferreira-Pinto[3], Rafael Massinga[2], José C. Ramalho[3,4] and Ana Ribeiro-Barros[3,4]*

1 TropiKMan Doctorate Programme, Nova School of Business and Economics, Universidade NOVA de Lisboa, Portugal

2 Divisão de Agricultura, Instituto Superior Politécnico de Manica (DivAG-ISPM), Moçambique

3 Plant Stress and Biodiversity Lab, Linking Landscape, Environment, LEAF, Instituto Superior de Agronomia (ISA), Universidade de Lisboa, Portugal

4 GeoBioTec, Faculdade de Ciências e Tecnologia, Universidade NOVA de Lisboa, Portugal

*Address all correspondence to: aribeiro@isa.ulisboa.pt

References

[1] Boukar O, Belko N, Chamarthi S, Togola A, Batieno J, Owusu E, et al. Cowpea (*Vigna unguiculata*): Genetics, genomics and breeding. Plant Breeding. 2018;**0**:1-10. DOI: 10.1111/pbr.12589

[2] Bashir M, Ahmad Z, Ghafoor A. Cowpea aphid-borne mosaic potyvirus: A review. International Journal of Pest Management. 2002;**48**:155-168. DOI: 10.1080/09670870110118722

[3] Peoples M, Brockwell J, Herridge D, Rochester I, Alves B, Urquiaga S, et al. The contributions of nitrogen-fixing crop legumes to the productivity of agricultural systems. Symbiosis. 2009;**48**:1-17. DOI: 10.1007/BF03179980

[4] Herridge DF, Peoples MB, Boddey RM. Global inputs of biological nitrogen fixation in agricultural systems. Plant and Soil. 2008;**311**:1-18. DOI: 10.1007/s11104-008-9668-3

[5] Bacci ML. A Concise History of World Population. West Sussex: John Wiley & Sons; 2017. p. 300. DOI: 10.1002/9781119406822

[6] Long SP, Marshall-Colon A, Zhu XG. Meeting the global food demand of the future by engineering crop photosynthesis and yield potential. Cell. 2015;**161**:56-66. DOI: 10.1016/j.cell.2015.03.019

[7] Edenhofer O, Pichs-Madruga R, Sokona Y, Farahani E, Kadner S, Seyboth K, et al. Climate change 2014: Mitigation of climate change. In: Contribution of Working Group III to the Fifth Assessment Report of the Intergovernmental Panel on Climate Change. Cambridge University Press; 2014. p. 1435. Available from: https://www.ipcc.ch/site/assets/uploads/2018/02/ipcc_wg3_ar5_full.pdf [Accessed: February 5, 2019]

[8] Ribeiro-Barros AI, Moura IM, Ramalho JC, Máguas-Hanson C, Ribeiro NS. The potential of tree and shrub legumes in agroforestry systems. In: Amanullah, editor. Nitrogen in Agriculture—Updates. Rijeka, Croatia: InTech; 2018. Available from https://www.intechopen.com/books/nitrogen-in-agriculture-updates/the-potential-of-tree-and-shrub-legumes-in-agroforestry-systems [Accessed: February 5, 2019]

[9] CSIRO. State of the Climate. 2018. Available from: https://www.csiro.au/en/Showcase/state-of-the-climate [Accessed: February 5, 2019]

[10] United Nations. Sustainable Development Goals. 2015. Available from: https://sustainabledevelopment.un.org/content/documents/21252030%20Agenda%20for%20Sustainable%20Development%20web.pdf [Accessed: February 5, 2019]

[11] Timko M, Singh BB. Cowpea, a multifunctional legume. In: Moore PH, Ming R, editors. Genomics of Tropical Plants. New York: Springer; 2008. pp. 227-258. DOI: 10.1007/978-0-387-71219-2_10

[12] Goncalves A, Goufo P, Barros A, Domínguez-Perles R, Trindade H, Rosa EA, et al. Cowpea (*Vigna unguiculata* L. Walp), a renewed multipurpose crop for a more sustainable agri-food system: Nutritional advantages and constraints. Journal of the Science of Food and Agriculture. 2016;**96**:2941-2951. DOI: 10.1002/jsfa.7644

[13] Ohler TA, Nielsen SS, Mitchell CA. Varying plant density and harvest time to optimize cowpea leaf yield and nutrient content. Horticultural Science. 1996;**31**:193-197

[14] Singh U, Singh B. Tropical grain legumes as important human foods. Economic Botany. 1992;**46**:310-321

[15] Schönfeldt HC, Gibson HN. Dietary protein quality and malnutrition in Africa. British Journal of Nutrition. 2012;**108**:69-76. DOI: 10.1017/S0007114512002553

[16] Jayathilake C, Visvanathan R, Deen A, Bangamuwage R, Jayawardana BC, Nammi S, et al. Cowpea: An overview on its nutritional facts and health benefits. Journal of the Science of Food and Agriculture. 2018;**98**:4793-4806. DOI: 10.1002/jsfa.9074

[17] Boukar O, Massawe F, Muranaka S, Franco J, Maziya-Dixon B, Singh B, et al. Evaluation of cowpea germplasm lines for protein and mineral concentrations in grains. Plant Genetic Resources. 2011;**9**:515-522. DOI: 10.1017/s1479262111000815

[18] Zahran HH. *Rhizobium*-legume symbiosis and nitrogen fixation under severe conditions and in an arid climate. Microbiology and Molecular Biology Reviews. 1999;**63**:968-989

[19] Quin FM. Introduction. In: Singh BB, Mohan Raj DR, Dashiell KE, Jackai LEN, editors. Advances in Cowpea Research. International Institute of Tropical Agriculture (IITA) and Japan International Research Center for Agricultural Sciences (JIRCAS); 1997. pp. ix-xv

[20] Carsky RJ, Vanlauwe B, Lyasse O. Cowpea rotation as a resource managemnet technology for cereal-based systems in the savannas of West Africa. In: Fatokum CA, Tarawali SA, Singh BB, Kormawa PM, Tamò M, editors. Challenges and Opportunities for Enhancing Sustainable Cowpea Production. International Institute of Tropical Agriculture (IITA); 2002. pp. 252-266

[21] Fatokum CA, Tarawali SA, Singh BB, Kormawa PM. Challenges and Opportunities for Enhancing Sustainable Cowpea Production. International Institute of Tropical Agriculture (IITA); 2002. p. 433

[22] Bado V, Bationo A, Cescas M. Assessment of cowpea and groundnut contributions to soil fertility and succeeding sorghum yields in the Guinean savannah zone of Burkina Faso (West Africa). Biology and Fertility of Soils. 2006;**43**:171-176. DOI: 10.1007/s00374-006-0076-7

[23] Omoigui L, Ekeuro GC, Kamara AY, Bello LL, Timko MP, Ogunwolu GO. New sources of aphids [*Aphis craccivora* (Koch)] resistance in cowpea germplasm using phenotypic and molecular marker approaches. Euphytica. 2017;**213**:178. DOI: 10.1007/s10681-017-1962-9

[24] Abudulai M, Kusi F, Seini SS, Seidu A, Nboyine JA, Larbi A. Effects of planting date, cultivar and insecticide spray application for the management of insect pests of cowpea in northern Ghana. Crop Protection. 2017;**100**:168-176. DOI: 10.1016/j.cropro.2017.07.005

[25] Souleymane A, Aken'Ova M, Fatokun C, Alabi Y. Screening for resistance to cowpea aphid (*Aphis craccivora Koch*) in wild and cultivated cowpea (*Vigna unguiculata* L. Walp.) accessions. International Journal of Science, Environment and Technology. 2013;**2**:611-621

[26] Parh IA. The effects of *Empoasca dolichi* Paoli (Hemiptera: Cicadellidae) on the performance and yield of two cowpea cultivars. Bulletin of Entomological Research. 1983;**73**:25-32. DOI: 10.1017/S0007485300013766

[27] Singh SR, Allen DJ. Cowpea Pests and Diseases. International Institute

of Tropical Agriculture (IITA); 1978. Available from: http://newint.iita.org/wp-content/uploads/2016/06/COWPEA_PESTS_AND_DISEASES.pdf [Accessed: February 5, 2019]

[28] Singh BB, Mohan-Raj D, Dashiell K, Jackai L, editors. Advances in Cowpea Research. Ibadan: International Institute of Tropical Agriculture (IITA); 1997. p. 375

[29] Asante SK, Tamò T, Jackai LEN. Integrated management of cowpea insect pests using elite cultivars, date of planting and minimum insecticide application. African Crop Science Journal. 2001;**9**:655-666. DOI: 10.4314/acsj.v9i4.27587

[30] Bamaiyi LJ, Onu I, Amatobi CI, Dike MC. Effect of *Callosobruchus maculatus* infestation on nutritional loss on stored cowpea grains. Archives of Phytopathology and Plant Protection. 2006;**39**:119-127. DOI: 10.1080/03235400500180743

[31] Kébé K, Alvarez N, Espíndola A, Justyl F, Olivieri I, Sembène M. Insights into the genetic structure of the cowpea pest *Callosobruchus maculatus* in Africa. Journal of Pest Science. 2016;**89**:449-458. DOI: 10.1007/s10340-015-0688-5

[32] Umeozor OC. Effect of the infection of *Callosobruchus maculatus* (fab.) on the weight loss of sored cowpea (*Vigna unguiculata* (L.) Walp). Journal of Applied Science and Environmental Management. 2005;**9**:169-172. Available from: http://www.bioline.org.br/pdf?ja05032 [Accessed: February 4, 2019]

[33] Annan B, Schafers G, Twingey W. Impact of density o*f Aphids craccivora* (Aphididae) on growth and yield of susceptible and resistant cowpea cultivars. Annals of Applied Biology. 1996;**128**:183-193. DOI: 10.1111/j.1744-7348.1996.tb07315.x

[34] Dugje LO, Omoigui LO, Ekeleme F, Kamara AY, Ajeigbe H. Farmers' Guide to Cowpea Production in West Africa. International Institute of Tropical Agriculture (IITA); 2009. p. 19. DOI: 10.13140/2.1.1597.5361

[35] Ujagir R, Byrne O. Insect pests and their management. In: Erskine W, Muehlbaue FJ, Sarker A, Sharma B, editors. The Lentil: Botany, Production and Uses. Wallingford: CABI; 2009. pp. 282-305. Available from: http://nsdl.niscair.res.in/jspui/bitstream/123456789/493/1/revised%20insect%20pest%20and%20their%20management.pdf [Accessed: February 5, 2019]

[36] Bett B, Gollash S, Moore A, James W, Armatrong J, Walsh T, et al. Transgenic cowpeas *(Vigna unguiculata L. Walp)* expressing *Bacillus thuringiensis* Vip3Ba protein are protected against the Maruca pod borer (*Maruca vitrata*). Plant Cell, Tissue and Organ Culture. 2017;**131**:335-345. DOI: 10.1007/s11240-017-1287-3

[37] Abutu A. Brighter Days for African Farmers—As Pod-Borer Resistant Cowpea Edge Towards Commercialisation. 2017. Available from: https://aatfnews.aatf-africa.org/?p=669 [Accessed: February 5, 2019]

[38] Klopez. Is Genetically Modified Cowpea Safe? 2009. Available from: http://r4dreview.iita.org/index.php/2009/03/08/640/ [Accessed: February 5, 2019]

[39] ACB. GM and Seed Industry Eye Africa's Lucrative Cowpea Seed Markets: The Political Economy of Cowpea in Nigeria, Burkina Faso, Ghana and Malawi. 2015. 38 p. Available from: https://acbio.org.za/wp-content/uploads/2015/07/GM-Cowpea-report.pdf [Accessed: February 5, 2019]

[40] News Ghana. Farmers Opt for GM Cowpea. 2017. Available from: http://b4fa.org/farmers-opt-gm-cowpea/ [Accessed: February 5, 2019]

[41] Williams RJ. Diseases of cowpea (*Vigna unguiculata* (L.) Walp.) in Nigeria. International Journal of Pest Management. 1975;**21**:253-267. DOI: 10.1080/09670877509411407

[42] Baker TA, Nielsen SS, Shade RE, Singh BB. Physical and chemical attributes of cowpea lines resistant and susceptible to *Callosobruchus maculatus* (F.) (Coleoptera: Bruchidae). Journal of Stored Products Research. 1989;**25**:1-8. DOI: 10.1016/0022-474X(89)90002-7

[43] Cruz LP, de Sá LFR, Santos LA, Gravina GA, Carvalho AO, Fernandes KVS, et al. Evaluation of resistance in different cowpea cultivars to *Callosobruchus maculatus* infestation 2015. Journal of Pest Science. 2016;**89**:117-128. DOI: 10.1007/s10340-015-0657-z

[44] Aboagye D, Darko JO, Banadda N. Comparative study of hermetic and non-hermetic storage on quality of cowpea in Ghana. Chemical and Biological Technologies in Agriculture. 2017;**4**:10. DOI: 10.1186/s40538-017-0091-y

[45] Baoua IB, Bakoye O, Amadou L, Murdock LL, Baributsa D. Performance of PICS bags under extreme conditions in the Sahel zone of Niger. Journal of Stored Products Research. 2018;**76**:96-101. DOI: 10.1016/j.jspr.2018.01.007

[46] Baributsa D, Baoua IB, Bakoye ON, Amadou L, Murdock LL. PICS bags safely store unshelled and shelled groundnuts in Niger. Journal of Stored Products Research. 2017;**72**:54-58. DOI: 10.1016/j.jspr.2017.03.007

[47] Agbicodo EM, Fatokun CA, Bandyopadhyay R, Wydra K, Diop NN, Muchero W, et al. Identification of markers associated with bacterial blight resistance loci in cowpea [*Vigna unguiculata* (L.) Walp.]. Euphytica. 2010;**175**:215-226. DOI: 10.1007/s10681-010-0164-5

[48] Olowe T, Dina SO, Oladiran AO, Olunuga BA. The control of weed, pest and disease complexes in cowpea (*Vigna unguiculata* (L.) Walp.) by the application of pesticides singly and in combination. Crop Protection. 1987;**6**:222-225. DOI: 10.1016/0261-2194(87)90042-1

[49] Adegbite AA, Amusa NA. The major economic field diseases of cowpea in the humid agro-ecologies of South-Western Nigeria. Archives of Phytopathology and Plant Protection. 2010;**43**:1608-1618

[50] Thottappilly G, Rossel HW. Virus diseases of cowpea in tropical Africa. Tropical Pest Management. 1992;**38**:337-348. DOI: 10.1080/09670879209371724

[51] Varma A. Sunn-Hemp mosaic virus. In: Van Regenmortel MHV, Fraenkel-Conrat H, editors. The Plant Viruses: The Rod-Shaped Plant Viruses. New York: Springer; 1986. pp. 249-266. DOI: 10.1007/978-1-4684-7026-0_13

[52] Silver S, Quan S, Deom CM. Completion of the nucleotide sequence of sunn-hemp mosaic virus: A tobamovirus pathogenic to legumes. Virus Genes. 1996;**13**:83-85. DOI: 10.1007/BF00576982

[53] Odedara O, Kumar L. Incidence and diversity of viruses in cowpeas and weeds in the unmanaged farming systems of savanna zones in Nigeria. Archives of Phytopathology and Plant Protection. 2016;**50**:1-12. DOI: 10.1080/03235408.2016.1241203

[54] Shoyinka SA, Thottappilly G, Adebayo GG, Anno-Nyako FO. Survey on cowpea virus incidence and distribution in Nigeria. International Journal of Pest Management. 1997;**43**:127-132. DOI: 10.1080/096708797228816

[55] Mukoye B, Mangeni B, Were H, Ndong'a M, Akech W. Occurrence

and biological characterization of cowpea aphid-borne mosaic virus infecting legumes in Western Kenya. In: Proceedings of the First Pan-African Congress on Knowledge Generation and Dissemination; 3-6 December 2017; Kenya. 2018

[56] Sarkar SC, Wang E, Wu S, Lei Z. Application of trap cropping as companion plants for the management of agricultural pests: A review. Insects. 2018;**9**:128. DOI: 10.3390/insects9040128

[57] Farooq M, Jabran K, Cheema ZA, Wahid A, Siddique KH. The role of allelopathy in agricultural pest management. Pest Management Science. 2011;**67**:493-506. DOI: 10.1002/ps.2091

[58] Ben-Issa R, Gomez L, Gautier H. Companion plants for aphid pest management. Insects. 2017;**8**:112. DOI: 10.3390/insects8040112

[59] Lehoczky E, Nelima MO, Szabó R, Szalai A, Nagy P. Allelopathic effect of Bromus spp. and Lolium spp. shoot extracts on some crops. Communications in Agricultural and Applied Biological Sciences. 2011;**76**:537-544

[60] Sing G, Sekhon HS. Integrated weed management in Pigeonpea [*Cajanus cajan* (L.) Millsp.]. World Journal of Agricultural Sciences. 2013;**9**:86-91. DOI: 10.5829/idosi.wjas.2013.9.1.27213

[61] Yadav T, Chopra NK, Yadav MR, Kumar R, Rathore DK, Soni PG, et al. Weed management in cowpea—A review. International Journal of Current Microbiology and Applied Sciences. 2017;**6**:1373-1385. DOI: 10.20546/ijcmas.2017.602.156

[62] Muranaka S, Fatokun C, Boukar OO. Stability of *Striga gesnerioides* resistance mechanism in cowpea under high-infestation level, low soil fertility and drought stresses. Journal of Food, Agriculture and Environment. 2011;**9**:313-318. Available from: https://cgspace.cgiar.org/handle/10568/42029

[63] Lado A, Umar SF, Usman YS, Kwalle KA. Efficacy of *Parkia biglobosa* fruit powder on the control of *Striga* in cowpea cropping systems in the Sudan-Savanna, Nigeria. Heliyon. 2018;**4**:e00733. DOI: 10.1016/j.heliyon.2018.e00733

[64] Gbègbèlègbè SD, Lowenberg-DeBoer J, Adeoti R, Lusk J, Coulibaly O. The estimated *ex ante* economic impact of Bt cowpea in Niger, Benin and northern Nigeria. Agricultural Economics. 2015;**46**:563-577. DOI: 10.1111/agec.12182

[65] Singh SK, Kakani VG, Surabhi GK, Reddy KR. Cowpea (*Vigna unguiculata* [L.] Walp.) genotypes response to multiple abiotic stresses. Journal of Photochemistry and Photobiology. 2010;**100**:135-146. DOI: 10.1016/j.jphotobiol.2010.05.013

[66] Craufurd PQ, Bojang M, Wheeler TR, Summerfield RJ. Heat tolerance in cowpea: Effect of timing and duration of heat stress. Annals of Applied Biology. 1998;**133**:257-267. Available from: https://onlinelibrary.wiley.com/doi/epdf/10.1111/j.1744-7348.1998.tb05826.x [Accessed: February 4, 2019]

[67] Iwuagwu M, Ogbonnaya CI, Onyike NB. Physiological response of cowpea [*Vigna unguiculata* (L.) Walp] to drought: The osmotic adjustment resistance strategy. Academic Journal of Science. 2017;7:329-344. Available from: file:///C:/Users/aribeiro/AppData/Local/Packages/Microsoft.MicrosoftEdge_8wekyb3d8bbwe/TempState/Downloads/MaryO.Iwuagwuetal%20(1).pdf [Accessed: February 4, 2019]

[68] Hall AE, Cisse N, Thiaw S, Elawad HOA, Ehlers JD, Ismail AM,

et al. Development of cowpea cultivars and germplasm by the bean/cowpea CRSP. Field Crops Research. 2003;**82**:103-134. DOI: 10.1016/S0378-4290(03)00033-9

[69] Ibrahim AR, Kiari SA, Mensah B, Akromah R. Water stress and water use efficiency in cowpea [*Vigna unguiculata* (L.) Walp.] under controlled environment. International Journal of Agricultural Science Research. 2013;**2**:191-199. Available from: file:///C:/Users/aribeiro/AppData/Local/Packages/Microsoft.MicrosoftEdge_8wekyb3d8bbwe/TempState/Downloads/WaterUseEffBioetMoi%20(1).pdf [Accessed: February 4, 2019]

[70] Jemo M, Sulieman S, Bekkaoui F, Olomide OAK, Hashem A, Abd Allah EF, et al. Comparative analysis of the combined effects of different water and phosphate levels on growth and biological nitrogen fixation of nine cowpea varieties. Frontiers in Plant Science. 2017;**8**:2111. DOI: 10.3389/fpls.2017.02111

[71] Agbicodo EM, Fatokun S, Muranaka R, Visser RG, Van Der Linden CGS. Breeding drought tolerant cowpea: Constraints, accomplishments, and future prospects. Euphytica. 2009;**167**:353-370. DOI: 10.1007/s10681-009-9893-8

[72] Adejumo TO, Florini DA, Ikotun T. Screening of cowpea cultivars for resistance to leaf smut. Crop Protection. 2001;**20**:303-309. DOI: 10.1016/S0261-2194(00)00155-1

[73] Adipala E, Takan JP, Mukalere Z. Preliminary evaluation of cowpea lines for resistance to zonate leaf spots and bacterial blight. East African Agricultural and Forestry Journal. 1995;**61**:55-61. DOI: 10.4314/eaafj.v61i1.46792

[74] Asare-bediako E, Vera EA, Aaron AT. Phenotypic and serological evaluation of cowpea (*Vigna unguiculata* L. Walp) genotypes for resistance to viral infection under field conditions. Journal of Plant Breeding and Crop Science. 2018;**10**:169-177

[75] Kareem K, Adegbite AA, Ayoola OT, Olayinka RB, Oloyede-Kamiyo QO. Evaluation of cowpea genotypes for infection by two aphid-borne viruses. World Rural Observations. 2016;**8**:80-88. Available from: http://www.sciencepub.net/rural/rural080416/13_31328wro080416_80_88.pdf [Accessed: February 4, 2019]

[76] Taiwo KA. The potential of cowpea as human food in Nigeria. Food Reviews International. 1998;**14**:351-370. DOI: 10.1080/87559129809541168

[77] Singh B. Cowpea Breeding at Iita: Highlights of Advances and Impacts. International Institute of Tropical Agriculture (IITA); 2018. 4 p. Available from: file:///C:/Users/aribeiro/AppData/Local/Packages/Microsoft.MicrosoftEdge_8wekyb3d8bbwe/TempState/Downloads/COWPEA_BREEDING_AT_IITA_HIGHLIGHTS_OF_ADVANCES_AND%20(1).pdf [Accessed: February 4, 2019]

[78] Karapanos I, Papandreou A, Skouloudi M, Makrogianni D, Fernández JA, Rosa E, et al. Cowpea fresh pods—A new legume for the market: Assessment of their quality and dietary characteristics of 37 cowpea accessions grown in southern Europe. Journal of Science and Food Agriculture. 2017;**97**:4343-4352. DOI: 10.1002/jsfa.8418

[79] Pottorff M, Ehlers JD, Fatokun C, Roberts PA, Close TJ. Leaf morphology in cowpea [*Vigna unguiculata* (L.) Walp]: QTL analysis, physical mapping and identifying a candidate gene using synteny with model legume species. BMC Genomics. 2012;**13**:234. DOI: 10.1186/1471-2164-13-234

[80] Boukar O, Fatokun CA, Roberts PA, Abberton M, Huynh BL, Close TJ, et al. Cowpea. In: Ron D, editor. Handbook of Plant Breeding. New York: Springer; 2015. pp. 219-250. DOI: 10.1007/978-1-4939-2797-5_7

[81] Georges F, Ray H. Genome editing of crops: A renewed opportunity for food security. GM Crops & Food. 2017;**8**:1-12. DOI: 10.1080/21645698.2016.1270489

[82] Poiré R, Chochois V, Sirault XR, Vogel JP, Watt M, Furbank RT. Digital imaging approaches for phenotyping whole plant nitrogen and phosphorus response in *Brachypodium distachyon*. Journal of Integrative Plant Biology. 2014;**56**:781-796. DOI: 10.1111/jipb.12198

[83] Akdemir D, Sanchez JI, Jannink JL. Optimization of genomic selection training populations with a genetic algorithm. Genetics, Selection, Evolution. 2015;**47**:38. DOI: 10.1186/s12711-015-0116-6

[84] Crain J, Mondal S, Rutkoski J, Singh RP, Poland J. Combining high-throughput phenotyping and genomic information to increase prediction and selection accuracy in wheat breeding. The Plant Genome. 2018;**11**:170043. DOI: 10.3835/plantgenome2017.05.0043

[85] Guo G, Zhao F, Wang Y, Zhang Y, Du L, Su G. Comparison of single-trait and multiple-trait genomic prediction models. BMC Genetics. 2014;**15**:30. DOI: 10.1186/1471-2156-15-30

[86] Hammer G, Messina C, van Oosterom E, Chapman S, Singh V, Borrell A, et al. Molecular breeding for complex adaptive traits: How integrating crop ecophysiology and modelling can enhance efficiency. In: Yin X, Struik P, editors. Crop Systems Biology. Cham: Springer; 2016. pp. 147-162. DOI: 10.1007/978-3-319-20562-5_7

[87] Hammer G. Molecular breeding for complex adaptive traits-can crop ecophysiology and modelling ease the pain? International Sugar Journal. 2014;**116**:64-68. Available from: file:///C:/Users/aribeiro/AppData/Local/Packages/Microsoft.MicrosoftEdge_8wekyb3d8bbwe/TempState/Downloads/HammerIntSugarJ2014%20(1).pdf [Accessed: February 4, 2019]

[88] Lado B, Vásquez D, Quincke M, Silva P, Aguilar I, Gutiérrez L. Resource allocation optimization with multi-trait genomic prediction for bread wheat (*Triticum aestivum* L.) baking quality. Theoretical and Applied Genetics. 2018;**131**:2719-2731. DOI: 10.1007/s00122-018-3186-3

[89] Van Eeuwijk FA, Bustos-Korts D, Millet EJ, Boer MP, Kruijer W, Thompson A, et al. Modelling strategies for assessing and increasing the effectiveness of new phenotyping techniques in plant breeding. Plant Science. 2018. DOI: 10.1016/j.plantsci.2018.06.018. In press

[90] Watson A, Ghosh S, Williams MJ, Cuddy WS, Simmonds J, Rey MD, et al. Speed breeding is a powerful tool to accelerate crop research and breeding. Nature Plants. 2018;**4**:23-29. DOI: 10.1038/s41477-017-0083-8

[91] O'Connor DJ, Wright GC, Dieters MJ, George DL, Hunter MN, Tatnell JR, et al. Development and application of speed breeding technologies in a commercial peanut breeding program. Peanut Science. 2013;**40**:107-114. DOI: 10.3146/PS12-12.1